SpringerBriefs in Mathematics

T0232295

SpringerBriefs in Mathematics showcases expositions in all areas of mathematics and applied mathematics. Manuscripts presenting new results or a single new result in a classical field, new field, or an emerging topic, applications, or bridges between new results and already published works, are encouraged. The series is intended for mathematicians and applied mathematicians.

For further volumes:
http://www.springer.com/series/10030

Katalin A. Bencsáth • Marianna C. Bonanome
Margaret H. Dean • Marcos Zyman

Lectures on Finitely Generated Solvable Groups

 Springer

Katalin A. Bencsáth
Department of Mathematics
  and Computer Science
Manhattan College
New York, NY, USA

Margaret H. Dean
Department of Mathematics
Borough of Manhattan Community College
The City University of New York
New York, NY, USA

Marianna C. Bonanome
Department of Applied Mathematics
  and Computer Science
New York City College of Technology
The City University of New York
Brooklyn, NY, USA

Marcos Zyman
Department of Mathematics
Borough of Manhattan Community College
The City University of New York
New York, NY, USA

ISSN 2191-8198   ISSN 2191-8201 (electronic)
ISBN 978-1-4614-5449-6   ISBN 978-1-4614-5450-2 (eBook)
DOI 10.1007/978-1-4614-5450-2
Springer New York Heidelberg Dordrecht London

Library of Congress Control Number: 2012945939

Mathematics Subject Classification (2010): 20-01, 20E07, 20E22, 20F16, 20C07, 20F05

Printed on acid-free paper

Springer is part of Springer Science+Business Media (www.springer.com)

For Gilbert

Foreword

When first encountered many students think infinite group theory is quite difficult. Certainly, it is not a breeze to understand. But this introduction gives graduate students and some undergraduates and perhaps others the opportunity to understand some aspects, important ideas and methods that are important in group theory and needed for further study. Exposition is measured and easy to understand with many examples designed to augment and illuminate the text. Readers will not be overwhelmed by an imposing volume of material as this is not the aim of such an introduction. What is possible here is that the reader will be able to digest the material at an individual pace. This book will also permit students to deepen their knowledge, understanding and enjoyment of infinite groups. The main focus here is on solvable groups, an area of group theory that has been neglected somewhat in recent years. It is a valuable introduction to group theory and will enable both students and seasoned professionals to use group theory in their studies or the field that they are mainly interested in.

It is my pleasure to write this foreword to a subject that I have been involved in for many years and that has given me a great deal of pleasure. I think that many scholars will enjoy reading it and will learn a great deal at the same time.

New York, NY, USA   Gilbert Baumslag

Preface

These lecture notes are based on a course entitled "Topics in Group Theory"[1] given by Gilbert Baumslag at the Graduate Center of CUNY during the Spring 2007 semester that focused on finitely generated solvable groups from the perspective of combinatorial group theory.

Among the major results highlighted by the course were the following:

- There are continuously many isomorphism classes of finitely generated 3-solvable groups, and the same is true for the (proper) subclass comprised of all the (finitely generated) center-by-metabelian groups (P. Hall, Theorem 1.1).
- Any extension G of a group A by a group H can be embedded in an unrestricted wreath product of A by H (Krasner and Kaloujnine, Theorem 4.3).
- If a finitely presented group G has an infinite cyclic quotient then either G is an ascending HNN-extension or else G contains a free group of rank 2 (Bieri and Strebel, Theorem 5.1).
- A finitely presented solvable group is either finite or else virtually an ascending HNN-extension of a finitely generated group (Bieri and Strebel, Theorem 5.2).
- Finitely generated metabelian groups satisfy the maximal condition for normal subgroups (P. Hall, Theorem 6.1).
- There are only countably many isomorphism classes of finitely generated metabelian groups (P. Hall, Theorem 6.2).
- Any finitely generated metabelian group can be embedded in a finitely presented metabelian group (G. Baumslag, Theorem 7.1).

Much of the course was devoted to building foundations and before studying techniques later utilized in proving the important structure and embedding theorems by Krasner and Kaloujnine [12], Bieri and Strebel [7], and Baumslag [3]. Among the tools included were semi-direct products, wreath products, HNN-extensions, representations; among the techniques employed in various settings were Cayley's

---

[1] http://math.gc.cuny.edu/courses/CourseDescriptions/CourseDescriptionMath87200.pdf

and Frobenius' permutation representations and Hall's module-over-group-ring approach to the normal subgroup structure of metabelian groups (see [8–10]).

In these notes, background material that goes beyond a first year graduate algebra course and some basic elements of combinatorial group theory (such as basis and rank of a free group) is developed, reviewed, and discussed as needed. Augmenting the more compact classroom lecture notes we include details of proofs, comments, and examples to a greater extent; and yet a number of proofs are still only sketched or hinted at, if not absent. The reader is encouraged to fully complete them.

Preparatory material takes up a sizable portion of the course. Deviating from the "live" version here, for clarity's sake, we decided to relegate some of the proof details to the appendices.

Several chapters and sections include concrete examples. More assistance from the relevant literature can also be found in the bibliography at the end of the document.

The authors claim full responsibility for any errors appearing in these notes.

New York, NY, USA   Katalin A. Bencsáth
Brooklyn, NY, USA   Marianna C. Bonanome
New York, NY, USA   Margaret H. Dean
New York, NY, USA   Marcos Zyman

Acknowledgments

This collaboration was born of a common desire not just to share some of the major results presented to us by Professor Baumslag but to also convey some of the joy, enthusiasm, and heart that he lends to all of his lectures. Throughout many years Professor Baumslag has given the members of this team (both individually and as a whole) generous encouragement which he delivered via his sharp wit and unique humor. His courses have been the inspiration and source of our lasting interest in combinatorial group theory, and we hope that this product of our efforts aptly reflects our enormous gratitude to him.

Thanks to Gretchen Ostheimer for her helpful discussions and suggestions. Our thanks also go to Anthony Weaver for access to his notes.

We would also like to thank our families and loved ones for their unwavering support throughout the production of these lecture notes. Without them, this collaboration would not have been possible. Thank you Mihaly, Gianfranco, Luca, Gianni, David, Adriana, and María.

Finally, we are grateful to our alma mater, the CUNY Graduate School and University Center, especially the Department of Mathematics, for the lasting hospitality that allowed us to produce this work.

Contents

Chapter 1
Preliminaries

1.1 Notation and Terminology

We begin with a description of our adopted notational conventions and terminology. In agreement with typical notational practice, if G is a group, $\zeta(G)$ denotes the center of G and $Aut(G)$ the group of all automorphisms of G. If $X \subseteq G$, $x, g \in G$, then x^g stands for $g^{-1}xg$, the *conjugate* of x by g, while $[x, g]$ denotes $x^{-1}g^{-1}xg$, the *commutator* of x and g. We write $gp(X)$, respectively $gp_G(X)$, for the smallest among the subgroups (respectively normal subgroups) of G that contain X. If a subgroup H of G coincides with $gp(X)$, we also say that H is generated by X as a subgroup of G. If H is a normal subgroup of a group G, and its subset Y has the property that $H = gp_G(Y)$, we will say that Y generates H as a normal subgroup, or that H is the *normal closure of* Y in G. If X and Y are non-empty subsets of the group G, then $[X, Y]$ is short for $gp([x, y] \mid x \in X, y \in Y)$. In particular, we denote by $[G, G]$ or G' the *commutator subgroup* (also called the first *derived group*) which is the (normal) subgroup in G generated by the set of all commutators of G.

Throughout these notes we use the following well-known commutator identities, which are true for any group G and all elements $w, x, y, z \in G$:

$$[x, y]^w = [x^w, y^w], \tag{1.1}$$

$$[xy, z] = [x, z]^y [y, z], \tag{1.2}$$

$$[x, yz] = [x, z][x, y]^z. \tag{1.3}$$

For non-empty sets Y and A, we denote the group of all permutations of Y by S_Y and the set of all functions from Y to A by A^Y.

The occasional notational complexity of the material in these notes makes it necessary to pay close attention to the context in which we "combine" functions. We have adopted the convention that where the binary operation is composition we keep to the "right-handed" custom of "argument followed by map"; in particular, we treat

K.A. Bencsáth et al., *Lectures on Finitely Generated Solvable Groups*,
SpringerBriefs in Mathematics, DOI 10.1007/978-1-4614-5450-2_1,
© Katalin A. Bencsath, Marianna C. Bonanome, Margaret H. Dean, Marcos Zyman 2013

permutations in this way. Where the binary operation is point-wise multiplication, we use notation according to the "left-handed" custom of "function name first on the left and argument follows (in parentheses)".

Replacing G by G' in the process of forming the commutator subgroup results in what is called the second derived subgroup of G, $G'' = \left[G', G' \right]$. Similarly, $G^n = \left[G^{n-1}, G^{n-1} \right]$. Thus, we arrive at a succession of non-increasing subgroups of G called *the derived series* of G:

$$G = G^0 \geq G' \geq \cdots G^n \geq \cdots$$

and refer to G^n as the *nth derived group* of G.

Definition 1.1. A group G is termed *solvable* if $G^l = 1$ for some positive integer l. The least such l is called the *solvability length* of G.

Throughout, frequently implicitly, we rely on the following important facts. As doing so is a good opportunity for reviewing some of the fundamental techniques of elementary group theory, the reader may want to furnish proofs.

Proposition 1.1. *Subgroups and quotient groups of a solvable group G are solvable, of solvability length not exceeding that of G.*

Proposition 1.2. *A non-abelian finitely generated free group is not solvable.*

Example 1.1. For any field k, the group of all invertible $n \times n$ matrices with entries from k contains the subgroup $T(n,k)$ of lower triangular matrices. $T(n,k)$ is a solvable group. If k is $\mathbb{Q}$, the field of rational numbers, and $l(n)$ denotes the solvability length of $T(n,\mathbb{Q})$, then $l(n) \to \infty$ as $n \to \infty$. Thus, these groups yield solvable groups of arbitrary solvability length.

In particular, $T(3,k)$ has solvability length 2. Groups of solvability length at most 2 are called *metabelian*. Given a metabelian group G, we have a short exact sequence

$$1 \to G' \to G \to G/G' \to 1$$

where both G' and G/G' are abelian, so G is an abelian group extended by an abelian group. The reader may wish to verify that, in fact, any group that is an extension of an abelian group by an abelian group is metabelian.

1.2   A Result on 3-Solvable Groups

A great portion of these notes is concerned with the relationship between finitely generated metabelian groups and finitely presented metabelian groups. As we will see (Theorem 6.2), the number of isomorphism classes of finitely generated metabelian groups is countably infinite.

While the change by just one in the solvability length is seemingly minor, the existence of a finitely generated 3-solvable group that contains a free abelian group of countably infinite rank as its center points to a sharply differing reality in the universe of finitely generated 3-solvable groups. This is underscored by the theorem below.

Theorem 1.1. *There exist continuously many non-isomorphic finitely generated 3-solvable groups.*

The proof of Theorem 1.1 follows from Proposition 1.1, along with the results presented in the remainder of this section.

We begin with the description of a two-generator matrix group. Let $\mathbb{Q}(x)$ denote the field of quotients of $\mathbb{Q}[x]$, the ring of polynomials in the single variable x with rational coefficients. Consider the matrices A and B with entries from $\mathbb{Q}(x)$:

$$A = \begin{pmatrix} 1 & 0 & 0 \\ 0 & x & 0 \\ 0 & 0 & 1 \end{pmatrix} \quad \text{and} \quad B = \begin{pmatrix} 1 & 0 & 0 \\ 1 & 1 & 0 \\ 0 & 1 & 1 \end{pmatrix}.$$

Appendix A contains an outline of steps establishing Lemma 1.1.

Lemma 1.1. *The center of $G = gp(A, B)$ is a free abelian group of countably infinite rank. Furthermore, $G'' \leq \zeta(G)$.*

The next component in the proof of Theorem 1.1 is the following:

Proposition 1.3. *A free abelian group A of infinite rank has continuously many different subgroups.*

Proof. Let $X = \{x_1, x_2, \ldots\}$ be a free basis for the free abelian group A. For each of the continuously many increasing sequences $\sigma = \sigma_1, \sigma_2, \ldots$ of positive integers, put

$$A_\sigma = gp(x_{\sigma_1}, x_{\sigma_2}, \ldots).$$

Then $A_\sigma = A_\tau$ if and only if $\sigma = \tau$, and that confirms the presence of continuously many distinct (though isomorphic) subgroups in A.   □

We end with a general theorem which is due to Jaques Lewin.

Theorem 1.2. *Suppose G is a finitely generated group with an uncountable family $\{C_\sigma\}$ of distinct normal subgroups of G. Then the quotients G/C_σ fall into uncountably many isomorphism classes.*

Proof. Consider the uncountable collection of quotient groups $\{G_\sigma = G/C_\sigma\}$, all finitely generated. Assume, contrary to the statement of the theorem, that these quotients fall into countably many isomorphism classes. This means that there is an uncountable set Φ and $\phi_0 \in \Phi$ such that $G_{\phi_0} \cong G_\phi$ for all $\phi \in \Phi$. If $v_\phi : G \to G_\phi$ is the canonical homomorphism, then for each isomorphism $\gamma_\phi : G_\phi \to G_{\phi_0}$, the map $\eta_\phi = v_\phi \gamma_\phi$ is a homomorphism from G to G_{ϕ_0} with kernel C_ϕ.

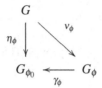

All these distinct normal subgroups C_ϕ would give us continuously many *distinct* homomorphisms η_ϕ from the finitely generated group G into the finitely generated group G_{ϕ_0}. However, the cardinality of the collection of these homomorphisms is bounded by the cardinality of all maps from a finite (generating) set of G into a countable set, a contradiction.   □

Regarding the matrix group $G = gp(A, B)$ of Lemma 1.1, we note that the fact that G'' lies in $\zeta(G)$ makes $G/\zeta(G)$ metabelian. This particular kind of 3-solvable group is called *center-by-metabelian* since it is an extension of a central subgroup by a metabelian group. G has continuously many non-isomorphic quotients, and these are also 3-solvable and center-by-metabelian.

Chapter 2
Tools: Presentations and Their Calculus

Let G be a group and X a set. A map $\vartheta : X \to G$ is called a *generation map* if $gp(X\vartheta) = G$. In such a case, ϑ extends to a homomorphism θ from the free group F on X onto G. This is known as the universal mapping property of free groups.

We write

$$G = \langle X; R \rangle$$

and term the right-hand side a *presentation* for G if the subset R of F generates the kernel of θ as a normal subgroup of F, that is, $ker\,\theta = gp_F(R)$. The group G is termed *finitely generated* if X can be chosen finite, *finitely presented* or *finitely presentable* if both X and R can be chosen finite. We also refer to θ as a *presentation map* (for G) and call R a set of *defining relators*.

We sometimes use the notation

$$G = \langle X; \{r = 1 \mid r \in R\} \rangle$$

in place of $G = \langle X; R \rangle$, and we refer to the expressions $r = 1$ $(r \in R)$ as *defining relations* for G.

The reader should note that there is a comparable notion of a relatively free group and corresponding relative presentation for certain classes of groups; such as, for example, the metabelian groups. If F is an absolutely free group on a set X, then F/F'' is metabelian and, within the class of metabelian groups, it possesses the universal mapping property. We call F/F'' the *free metabelian group* on X. Every metabelian group is a quotient of a free metabelian group.

The following observation appears quite simple, yet it is a surprisingly handy tool for dealing with groups given by generators and defining relations.

Lemma 2.1 (von Dyck). *Let* $X = \{x_i \mid i \in I\}$ *and suppose that*

$$G = \langle X; R \rangle.$$

K.A. Bencsáth et al., *Lectures on Finitely Generated Solvable Groups*,
SpringerBriefs in Mathematics, DOI 10.1007/978-1-4614-5450-2_2,
© Katalin A. Bencsath, Marianna C. Bonanome, Margaret H. Dean, Marcos Zyman 2013

Let $\theta : X \to G$ be the map that comes with this presentation, and put $\theta(x_i) = y_i \in G$. Let $Y = \{y_i \mid i \in I\}$, H another group, and $\phi : Y \to H$ a set map. If for every word $r = (x_{i_1}, \ldots, x_{i_n}) \in R$,

$$r(\phi(y_{i_1}), \ldots, \phi(y_{i_n})) = 1$$

in H, then ϕ can be extended to a homomorphism φ from G to H.

Due to its utility, von Dyck's valuable lemma is used repeatedly throughout these notes, often without explicit mention. We encourage the reader to recognize all those instances.

Whenever we deal with groups in terms of presentations, we bear in mind that writing

$$G = \langle X; R \rangle$$

assumes a presentation map θ and a homomorphism from the free group $F = F(X)$ onto the group G.

Different-looking presentations may well be describing isomorphic groups. In particular, a finitely presented group may have other presentations that need not use finitely many relators or finitely many generators. It can be convenient to be able to switch between presentations of the same group. The following theorem and lemma dealing with the "calculus of presentations" describe some "rules" for permissible manipulations that leave the group of the presentation unchanged (up to isomorphism). Both are important results for finitely presentable groups.

Theorem 2.1 (Tietze).

- **T1** Suppose that $G = \langle X; R \rangle$. Let Y be a set, disjoint from X, and consider a set of X-words w_y for each $y \in Y$. Then

$$G = \left\langle X \cup Y; R \cup y^{-1} w_y \ (y \in Y) \right\rangle.$$

- **T1′** Suppose that $G = \left\langle X \cup Y; R \cup y^{-1} w_y \ (y \in Y) \right\rangle$, where Y is a set disjoint from X, each w_y is an X-word, and the elements of R are X-words. Then

$$G = \langle X; R \rangle.$$

- **T2** Suppose that $G = \langle X; R \rangle$ and $S \subseteq gp_F(R)$ in the free group F on X. Then

$$G = \langle X; R \cup S \rangle.$$

- **T2′** Suppose that $G = \langle X; R \cup S \rangle$, where $S \subseteq gp_F(R)$ in the free group F on X. Then

$$G = \langle X; R \rangle.$$

The moves Theorem 2.1 specifies are called Tietze transformations. Appendix B contains the proof of Tietze's theorem.

Lemma 2.2 (B. H. Neumann). *Suppose the group G is known to be finitely presentable and a presentation*

$$G = \langle x_1, \ldots, x_m; r_1, r_2, \ldots \rangle$$

with a finite set of generators and a (possibly infinite) countable set of relators is given for G. Then for some positive integer n, G can also be presented by the finite presentation

$$G = \langle x_1, \ldots, x_m; r_{i_1}, r_{i_2}, \ldots, r_{i_n} \rangle.$$

In other words, if a group G is finitely presentable, then in every presentation of G that involves a finite set of generators, some finite subset of the given relators will suffice to present G.

The proof is a straightforward application of Tietze transformations.

Example 2.1. To illustrate the difficulty of recognizing isomorphic groups, we offer a presentation of a group G that turns out to be a trivial group:

$$G = \langle a, b; a^{-1}ba = b^2; b^{-1}ab = a^2 \rangle.$$

To see that G is trivial, consider $a^{-1}ba = b^2$. This leads to $b^{-1}a^{-1}ba = b$. Now $b^{-1}a^{-1}b = (b^{-1}ab)^{-1}$ which implies $a^{-2}a = b$ and $a^{-1} = b$. Next, substituting into $bba = b^2$ results in $a = 1$ and so $a^{-1} = 1$ which implies $b = 1$.

The natural question one can ask is whether there is an algorithm deciding if any given presentation represents the trivial group. In fact, this triviality problem is algorithmically undecidable.

Chapter 3
Constructions

3.1 Semi-direct Products

3.1.1 Internal

Suppose G is a group with subgroups T and U such that

(1) $G = TU$
(2) $U \trianglelefteq G$
(3) $T \cap U = \{1\}$

In these circumstances, every nontrivial element $g \in G$ can be written uniquely as a product $g = tu$ ($t \in T$, $u \in U$) since $g = tu = t'u'$ is equivalent to $u(u')^{-1} = t^{-1}t'$. For $t \in T$, let $\hat{t}$ denote the map $u \mapsto t^{-1}ut$ ($u \in U$), an automorphism of U. We find a "rule" for combining these normal forms through a brief calculation: If $g_1 = t_1 u_1$ and $g_2 = t_2 u_2$, then

$$t_1 u_1 t_2 u_2 = t_1 (t_2 t_2^{-1}) u_1 t_2 u_2 = t_1 t_2 (t_2^{-1} u_1 t_2) u_2 = t_1 t_2 (u_1)^{\hat{t_2}} u_2.$$

The group G that satisfies the conditions (1)–(3) above is called the (internal) semi-direct product (of its subgroups T and U), and we write $G = U \rtimes T$. Note that the map $\varphi : T \to Aut(U)$ defined by $t\varphi = \hat{t}$ is a homomorphism.

In the special case when $\hat{t}$ is the identity automorphism of U for every $t \in T$, we have $G = U \times T$.

3.1.2 External

The description of an internal semi-direct product suggests a more general construction of a new group out of a suitably linked pair of groups. The new group, their

K.A. Bencsáth et al., *Lectures on Finitely Generated Solvable Groups*,
SpringerBriefs in Mathematics, DOI 10.1007/978-1-4614-5450-2_3,
© Katalin A. Bencsath, Marianna C. Bonanome, Margaret H. Dean, Marcos Zyman 2013

external semi-direct product, turns out to be internally the semi-direct product of its subgroups isomorphic to the starting pair of groups. The specifics are as follows. Let U and T be a pair of groups and suppose that φ is a homomorphism from T to $Aut(U)$. The Cartesian set product $\{(t,u) \mid t \in T, u \in U\}$ becomes a group under the operation $(t,u)(t',u') = (tt', u(t'\varphi)u')$. This group G is called the *semi-direct product* of U by T via φ, and we use the notation $G = U \rtimes_\varphi T$, with φ suppressed when it is clearly implied from the context. We say the homomorphism φ defines an *action* of T on U; if φ is a monomorphism, we say T acts *faithfully* on U.

G contains "natural" isomorphic copies $\overline{T} = \{(t,1) \mid t \in T\}$ of T and $\overline{U} = \{(1,u) \mid u \in U\}$ of U as trivially intersecting subgroups. It can be readily verified that

$$(t,1)^{-1}(1,u)(t,1) = (1, u(t\varphi)),$$

so $\overline{U} \trianglelefteq G$ and conjugation by the elements of $\overline{T}$ "realizes" the homomorphism $\varphi : T \to Aut(U)$. One noteworthy consequence is that if T acts faithfully on U in $G = U \rtimes_\varphi T$, then $T \cap \zeta(G) = 1$.

3.1.3 Examples

In each of the following examples, Λ is a unital ring (i.e., a ring with unity), Λ^+ the additive group of Λ, and A an isomorphic copy of Λ^+, written multiplicatively, where the unity element $1 \in \Lambda$ has the isomorphic image $a \in A$. If $I(\Lambda)$ is the group of units of Λ and $J \leq I(\Lambda)$, then for every $\alpha \in J$, right multiplication of the elements of Λ^+ by α defines an automorphism of Λ^+. Correspondingly, the mapping $\overline{\alpha} : g \mapsto g^\alpha$ $(g \in A)$ defines an automorphism of A, and the mapping from J to $Aut(A)$ given by

$$\mu : \alpha \mapsto \overline{\alpha}$$

is an (injective) homomorphism.

Suppose now that J is the image of some group T under a homomorphism ν. Then $\varphi = \nu\mu$ is a homomorphism from T to $Aut(A)$, and we can form the semi-direct product $A \rtimes_\varphi T$.

Example 3.1. Let $\Lambda = \mathbb{Z}$, and $T = \langle t; t^2 \rangle$ be the cyclic group of order 2. We have $I(\Lambda) = \{-1, 1\}$. In this case, Λ^+ is the additive group of $\mathbb{Z}$. Consider the homomorphism φ from T to $Aut(\mathbb{Z})$.

$$t\varphi : r \mapsto r(-1) \quad (r \in \mathbb{Z}).$$

A is now the multiplicative infinite cyclic group generated by a, the isomorphic image of the unity element 1 of $\mathbb{Z}$. The semi-direct product of A and T thus has the presentation

$$G = A \rtimes_\varphi T = \langle t, a; t^2 = 1, t^{-1}at = a^{-1} \rangle.$$

Example 3.2. Let $\Lambda = \mathbb{Z}[1/2]$, the ring of integers with $1/2$ adjoined, and $T = \langle t \rangle$, the infinite cyclic group. Here, Λ^+ is not finitely generated (Note that, like $\mathbb{Q}$, Λ^+ is torsion-free but not free abelian.). Now, $1/2$ is a unit in Λ. Mapping t to $1/2$ results in the homomorphism $\varphi : T \to Aut(\Lambda^+)$ that sends t to the "multiply-by-1/2" automorphism, $r \mapsto r\left(\frac{1}{2}\right)$ $(r \in \Lambda^+)$. The effect of that automorphism on the generator a (the image of the unity in the multiplicative version of Λ^+) is described by $t^{-1}at = a^{1/2}$, which leads to the following (awkward) presentation for the semi-direct product:

$$G = A \rtimes_\varphi T = \left\langle t, a; \ t^{-1}at = a^{1/2}, \ \left[a, a^{t^n}\right] = 1 \ (n \in \mathbb{Z}) \right\rangle.$$

Since v is a homomorphism, $tv = 1/2$ if and only if $t^{-1}v = 2$. The relation $t^{-1}at = a^{1/2}$ turns into $tat^{-1} = a^2$. Moreover, since

$$\left[a, a^{t^n}\right] = \left[a^{t^{-n}}, a\right]^{t^n} = \left[a^{2^n}, a\right]^{t^n} = 1 \ (n \in \mathbb{Z}),$$

the commuting relations are not really needed. Thus, we arrive at the more standard (finite) presentation

$$G = A \rtimes_\varphi T = \left\langle t, a; \ tat^{-1} = a^2 \right\rangle.$$

Example 3.3. Let $\Lambda = \mathbb{Z}[1/6]$ and $T = \langle t \rangle$, the infinite cyclic group. Note that $2/3 \in I(\Lambda)$. Mapping t to $2/3$ gives rise to the homomorphism φ from T to $Aut(A)$ described by

$$t\varphi : g \mapsto g^{2/3} \ (g \in A).$$

As before, the resulting presentation

$$G = A \rtimes_\varphi T = \left\langle t, a; \ t^{-1}at = a^{2/3}, \ \left[a, a^{t^n}\right] \ (n \in \mathbb{Z}) \right\rangle$$

can be converted to

$$G = \left\langle t, a; \ t^{-1}a^3t = a^2, \ \left[a, a^{t^n}\right] \ (n \in \mathbb{Z}) \right\rangle$$

using Tietze transformations (see Theorem 2.1). In this case, however, the infinitely many commuting relations cannot be dispensed with, as we shall see later in Example 5.1.

Example 3.4. Let $\Lambda = \mathbb{Z}[x, x^{-1}]$, the ring of integers with x and x^{-1} adjoined, and $T = \langle t \rangle$, the infinite cyclic group. Then $x \in I(\Lambda)$. Λ^+ is the additive group of polynomials in x and x^{-1}. Let a_i denote the image of x^i under the isomorphism from Λ^+ to A, with $a = a_0$ the image of 1. $v : T \to I(\Lambda)$ defined by mapping t to x gives rise to the induced homomorphism φ from T to $Aut(A)$ described by

$$t\varphi : a_i \mapsto a_{i+1} \ (i \in \mathbb{Z}).$$

Notice that A is free abelian of infinite rank, generated by the set $\left\{a^{t^i}\right\}$, or $\{a_i\}$ $(i \in \mathbb{Z})$. The action of t on the generators of A is:

$$t^{-1}a_i t = a_i^t = a_{i+1}.$$

A presentation for A is $\langle \ldots, a_{-1}, a_0, a_1, \ldots; [a_i, a_j] \ (i,j) \in \mathbb{Z} \rangle$. A presentation for the semi-direct product of A with T via φ is

$$G = A \rtimes_\varphi T = \left\langle a, t; \left[a, a^{t^i}\right] (i \in \mathbb{Z}) \right\rangle.$$

Although G is generated by two elements, its subgroup A is not finitely generated. In the next section, we will see this group again as an example of a so-called wreath product.

3.2   Wreath Products

3.2.1   Internal and Restricted

Suppose that A and T are subgroups of a group W. Put $B = gp_W(A)$ and, for $s \in T$, denote $s^{-1}As$ by A^s. W is called the *(restricted, or standard) wreath product* of A by T if

(1) $W = gp(A, T)$
(2) $B = \prod_{s \in T} A^s$

In such a case we call A the bottom group, B the base group and T the top group of the wreath product W and use the notation $W = A \wr T$.

As conjugates of A by distinct elements of T commute, it also follows that, in $W = A \wr T$, $B \cap T = 1$, for $t \in B \cap T$ yields $t = a_1^{t_1} \cdots a_n^{t_n}$ $(n \geq 1)$ with $a_i \in A$, $t_i \in T$ $(i = 1, \cdots, n)$; whence

$$A^t = A^{a_1^{t_1} \cdots a_n^{t_n}} = A = A^1,$$

thus, $t = 1$.

For $t \in T$, the mapping $\hat{t} : A^s \to A^{st}$ $(s \in T)$ is an automorphism of B, and $\varphi : T \to Aut(B)$ defined by $t\varphi = \hat{t}$ is a homomorphism. Therefore, the wreath product is also a semi-direct product (of a special kind): $W = B \rtimes_\varphi T$, with T acting faithfully on B by permuting the direct factor copies of A in B according to the right regular representation of T (which we will discuss in the next chapter; cf. Sect. 4.1).

3.2.2 External

Guided by the previous analysis of the action of the top group on the base group in a wreath product, in this section we demonstrate how to construct wreath products for all pairs of groups.

Suppose that A and T are a given pair of groups. For each $s \in T$, let $A_s = \{a_s \mid a \in A\}$ be an isomorphic copy of A. Form $B = \prod_{s \in T} A_s$.

For $t \in T$, define $\hat{t} \in Aut(B)$ by $a_s \hat{t} = a_{st}$ $(a_s \in A_s, s \in T)$. Routine checks show that $\hat{t}$ is an automorphism of B that permutes the copies of A in B. Since the mapping $\varphi : t \to \hat{t}$ $(t \in T)$ is a homomorphism from T to $Aut(B)$, we can form the semi-direct product $W = B \rtimes_\varphi T$. Identifying T and B with their natural images in W, we know $W = TB$, and we can identify A with A_1, via φ. Then $a\hat{t} = a_1\hat{t} = a_t$ for all $a \in A$, so $A_t = A^{\hat{t}}$, whence $W = gp(A, T) = A \wr T$.

Actually, we can carry out a related construction in greater generality, for groups acting on sets.

Recall that if Y is a non-empty set and A is any group, we write A^Y for the set $\{\alpha : Y \to A\}$ of all functions from Y to A.

For example, when $Y = \{1, 2\}$ then $A^Y = \{(a_1, a_2) \mid a_i \in A \ (i = 1, 2)\}$; when $Y = \mathbb{N}$, then $A^Y = \{(a_1, a_2, \ldots) \mid a_i \in A\}$.

A^Y is a group under point-wise multiplication: $(\alpha\gamma)(y) = \alpha(y)\,\gamma(y)$ for $\alpha, \gamma \in A^Y$, or, using componentwise notation, if $\alpha = (a_1, a_2, \ldots)$, and $\gamma = (b_1, b_2, \ldots)$, then $\alpha\gamma = (a_1 b_1, a_2 b_2, \ldots)$. We call this group the *unrestricted direct product* of $|Y|$ copies of A and denote it by $A^Y = \prod_{y \in Y} A$.

Suppose now that T is a group acting on Y; that is, T comes with a homomorphism $\beta : T \to S_Y$. As $t\beta$ permutes Y, $t\beta$ induces a permutation of A^Y. This gives rise to a homomorphism $\varphi_\beta : T \to Aut(A^Y)$ that we define as follows, writing t for $t\beta$ and φ for φ_β.

$$\alpha \, t\varphi \, (y) = \alpha(yt^{-1}) \ (y \in Y, t \in T, \ \alpha \in A^Y). \tag{3.1}$$

To see that φ is indeed a homomorphism from T to $Aut(A^Y)$, observe that

$$\begin{aligned}
[(\alpha \, t_1 \varphi) \, t_2 \varphi](y) &= (\alpha \, t_1 \varphi)\left(yt_2^{-1}\right) \\
&= \alpha\left(yt_2^{-1}t_1^{-1}\right) \\
&= \alpha\left(y(t_1 t_2)^{-1}\right) \\
&= [\alpha(t_1 t_2 \varphi)](y). \tag{3.2}
\end{aligned}$$

Opting for t, rather than t^{-1}, on the right side of (3.1) above may appear more natural. However, the computation in (3.2) affirms that t^{-1} is needed if φ is to be a homomorphism; the use of t in place of said t^{-1} would result in an *anti-homomorphism* φ, with $(t_1 \varphi)(t_2 \varphi) = (t_2 t_1)\varphi$.

The semi-direct product $A^Y \rtimes_\varphi T$ constructed with the action of T on A^Y defined above is referred to as the *unrestricted wreath product* of A by T and denoted by $\overline{W} = A \bar{\wr}_Y T$ (suppressing reference to the actions β and φ).

The subgroup $\{f : Y \to A \mid f$ has finite support in $Y\}$ of A^Y, denoted by $A^{(Y)}$, is called the *restricted direct product* of copies of A indexed by the elements of Y. Clearly if $|Y| < \infty$, then $A^Y = A^{(Y)}$.

It is worth noting that in the group $W = A \wr T$ discussed in Sect. 3.2.1, the subgroup B has the alternate description

$$B = \{f : T \to A \mid f \text{ has finite support in } T\},$$

so $B = A^{(T)}$.

In these terms, the conjugation action of T on B in Sect. 3.2.1 is described as follows: if $f \in B$, $s, t \in T$, then

$$f^{\hat{t}}(s) = f(st^{-1}) \ (s \in T),$$

where $f^{\hat{t}}$ stands for the function in $B = A^{(T)}$ to which $t \in T$ carries $f \in A^{(T)}$ by conjugation.

For the subgroup $W = gp(A^{(Y)}, T)$ in $\overline{W}$, we can verify that $W = A^{(Y)} \rtimes_\varphi T$ since $A^{(Y)} \leq A^Y$ and $A^{(Y)}$ is closed under the action of T. Accordingly, we denote this subgroup W of $\overline{W}$ by $A \wr_Y T$, which is known as the *restricted (or standard) wreath product* of A by T.

3.2.3   Examples

Example 3.5. In the following example, we view our wreath product internally. Let $A = \langle a \rangle$ and $T = \langle t \rangle$ and put $A^{t^i} = gp(a^{t^i}) = A_i$. For ease of notation, let $a_i = a^{t^i}$. $B = \prod_{i \in \mathbb{Z}} A_i$ is a free abelian group freely generated by the set $\{\cdots, a_{-1}, a_0, a_1, \cdots\}$. Our wreath product in this case, $W = A \wr T = B \rtimes T = gp(B, t) = gp(a_0, t) = gp(a, t)$, is exactly the group of Example 3.4, Sect. 3.1.3. It is a wreath product of an infinite cyclic group by an infinite cyclic group, and a presentation for it is

$$\left\langle a, t; \left[a, a^{t^j}\right] = 1 \ (j \in \mathbb{Z}) \right\rangle.$$

Example 3.6. In this example, we build an external wreath product from the following ingredients:

- $Y = \{1, 2, 3\}$
- $T = \langle a, b; \ a^2 = b^2 = (ab)^2 = 1 \rangle = \{1, a, b, ab\}$ (the Klein-4 group)
- $A = \mathbb{Z} \Rightarrow A^Y = \mathbb{Z} \times \mathbb{Z} \times \mathbb{Z}$

Then $S_Y = S_3 = \langle x = (1\ 2), w = (1\ 2\ 3); \ x^2 = w^3 = 1, wx = xw^2 \rangle$. These ingredients allow for an extremely limited number of choices for the homomorphism β;

we choose $a \mapsto x$ and $b \mapsto 1$. Adaptation of (3.1) from Sect. 3.2.2, leads to the action φ of T on A^Y. Written out in detail, the action is:

$$\alpha 1 \varphi(y) = \alpha(y),$$
$$\alpha a \varphi(y) = \alpha \left(yx^{-1}\right) = \alpha(yx) \ (x \text{ is a permutation of order 2}),$$
$$\alpha b \varphi(y) = \alpha(y),$$
$$\alpha(ab)\varphi(y) = \alpha(yx).$$

Now we form the semi-direct product $\overline{W} = W = A^Y \rtimes_\varphi T = A \overline{\wr}_Y T$. The reader may wish to try and write a presentation for this group W using knowledge about its structure.

Remark 3.1. To a person with no prior knowledge of the actual construction(s), the presentation found by the reader may not indicate the standard wreath product's way of a bottom group, base group, and top group working together. The group of said presentation would likely be interpreted as having the structure of

$$W = \left(\mathbb{Z} \wr \left\langle a; a^2 \right\rangle\right) \times \mathbb{Z} \times \left\langle b; b^2 \right\rangle.$$

3.3 *HNN*-Extensions

3.3.1 *Internal*

Suppose that G is a group, B is a subgroup of G, and t an element of G such that

$$G = gp(B, t).$$

Suppose, in addition, that H, K are subgroups of B, with

$$t^{-1}Ht = K.$$

Since every element of G is a product of various powers of t and elements of the subgroup B, a typical element $g \in G$ can be expressed as a formal product

$$g = t^{\varepsilon_1} b_1 \cdots t^{\varepsilon_n} b_n$$

where $b_i \in B$ (possibly 1) and $\varepsilon_i = \pm 1$.

Definition 3.1. A subword of the form

$$t^{\varepsilon_i} b_i t^{\varepsilon_{i+1}}$$

is called a *pinch* if $\varepsilon_i = -1$, $\varepsilon_{i+1} = 1$, and $b_i \in H$; or $\varepsilon_i = 1$, $\varepsilon_{i+1} = -1$, and $b_i \in K$.

Suppose that whenever a formal product $t^{\varepsilon_1} b_1 \cdots t^{\varepsilon_n} b_n$ turns out to equal 1 in G, it is guaranteed to contain a pinch. Then G is called an *HNN-extension* of the group B (called the *base group*) with *stable letter t* and *associated subgroups H and K*. If either H or K is the group B itself, we call G an *ascending HNN-extension*. When B, H, and K are known, we can write $G = \langle B, t; H^t = K \rangle$.

3.3.2 External

Given a group B and an isomorphism φ between its subgroups H and K, an *HNN-extension E of B* (with *associating isomorphism φ*) can be constructed by joining a new generator t that realizes φ as the effect of conjugation by t in the group E. Just as intuition might suggest, E can be described conveniently in terms of generators and relations as follows.

Suppose $B = \langle X; R \rangle$. Then for some index set I, there are X-words u_i, v_i so that $v_i = u_i \varphi$ $(i \in I)$, $H = gp(u_i \mid i \in I)$; therefore, $K = gp(v_i \mid i \in I)$. A presentation for E is

$$E = \left\langle X, t; R \cup \{ t^{-1} u_i t = v_i \mid i \in I \} \right\rangle,$$

where the generator t is a formal symbol not contained in X.

Indeed, this group E has the "naturally expected" properties:

1. B has an injected copy in this new group.
2. If a group like G of Sect. 3.3.1 is revealed to be an internal *HNN-* extension of B with one stable letter conjugating a pair of isomorphic subgroups of B to each other, then the presentation for E above also presents G (see Appendix C).

3.3.3 Examples

The group described in the first example, a *Baumslag-Solitar group*, is built with the method described in the previous section. It belongs to the family of two-generator, one-relator groups known as $BS(m, n)$. For each pair of nonzero integers m, n, $BS(m, n)$ is the group with presentation

$$\langle a, b; b a^m b^{-1} = a^n \rangle .$$

Many Baumslag-Solitar groups, including the one below, possess a remarkable property, which will be demonstrated here.

Example 3.7. Consider

- $B = \langle b \rangle$
- $H = gp(b^2)$
- $K = gp(b^3)$

and let the isomorphism $\varphi : H \to K$ be given by $b^2 \varphi = b^3$. The resulting group

$$E = \langle B,t;\ t^{-1}b^2t = b^3 \rangle = \langle b,t;\ t^{-1}b^2t = b^3 \rangle$$

is the Baumslag-Solitar group, $BS(2,3)$.

Now, define a map $\widehat{\alpha}$ on the generators of E where $b\widehat{\alpha} = b^2$ and $t\widehat{\alpha} = t$. $\widehat{\alpha}$ can be continued to a homomorphism $\alpha : E \to E$ which turns out to be an epimorphism. Therefore, $E \cong E/ker\alpha$. Yet the kernel of α is nontrivial, as we can see by considering $e\alpha$, where

$$e = [t^{-1}bt,b] = t^{-1}b^{-1}tb^{-1}t^{-1}btb.$$

Clearly $e \neq 1$ because it does not contain any pinches. However,

$$e\alpha = [(t^{-1}bt)\alpha,b\alpha] = [t^{-1}b^2t,b^2] = [b^3,b^2] = 1.$$

So E is isomorphic to one of its proper quotient groups. This is an example of a so-called *non-Hopfian* group.

Remark 3.2. It is also true that for any values of n,m other than $|m| = 1$ or $|n| = 1$ such that $\pi(m) = \pi(n)$ (where $\pi(m)$ denotes the set of prime divisors of m), the group $BS(m,n)$ is non-Hopfian.

Following are two examples of ascending *HNN*-extensions.

Example 3.8.
$$H = B = K = \langle b \rangle$$

B is a finitely presented abelian group. If both H and K coincide with B, then t acts as an automorphism of B. The only possibilities are

$$G = \langle b,t;\ t^{-1}bt = b \rangle = \mathbb{Z} \times \mathbb{Z}$$

or
$$G = \langle b,t;\ t^{-1}bt = b^{-1} \rangle = \mathbb{Z} \rtimes \mathbb{Z} \quad \text{(the infinite dihedral group)}.$$

Example 3.9. Ingredients:

- $B = \langle b \rangle$
- $H = gp(b^2)$
- $K = gp(b) = B$
- $\varphi : b^2 \mapsto b$

The group
$$E = \langle B,t;\ t^{-1}b^2t = b \rangle$$

is an ascending *HNN*-extension.

Putting
$$b = b_0, \; t^{-1}bt = b_1, \ldots, \; t^{-i}bt^i = b_i, \ldots,$$

and $gp_E(B) = gp(b_0, b_1, \ldots)$, we find that:

$$b_1^2 = b_0, \; b_2^2 = b_1, \ldots$$

and

$$B = gp(b_0) < gp(b_1) < gp(b_2) < \cdots < \bigcup_{i=0}^{\infty} t^{-i}Bt^i = gp_E(B).$$

Example 3.10 (Iterated HNN-extension). Let $E_1 = \langle b, t; \; b^t = b^3 \rangle$. E_1 is an ascending *HNN*-extension, with $B = \langle b \rangle$, $H = B$, $K = \langle b^3 \rangle$. We can add another stable letter to E_1 as follows.

Let $E_2 = \langle E_1, s; \; b^s = b^7, t^s = t \rangle$.

If this procedure is iterated n times, we have effectively added n commuting stable letters to our original base group. Of course, there are consequences to the new relations; such as, in E_2, we have $b^s b^{-2t} = b$.

Remark 3.3. There is another procedure known as a *generalized HNN*-extension, in which n stable letters are added simultaneously. In this case, the stable letters form a free subgroup of rank n of the *HNN*-extension.

Chapter 4
Representations and a Theorem of Krasner and Kaloujnine

4.1 Cayley's Permutation Representation

Sometimes we term a homomorphism θ from a group G into a group H a *representation* of G. In particular, if H is a group of matrices, we say that θ is a *matrix representation* of G, and if H is a subgroup of the symmetric group S_X for some set X, then we term θ a *permutation representation* of G. If the given homomorphism θ is one-to-one, the representation is termed *faithful*.

Theorem 4.1 (Cayley). *Every group G has a faithful permutation representation.*

Proof. Consider S_G the symmetric group on the set (underlying the group) G.

First, observe that right multiplying all the elements $x \in G$ by a fixed element $g \in G$ permutes the elements of G. Next, define a map

$$\sigma : G \to S_G$$

by assigning to each $g \in G$ its afore-mentioned right-multiplier map:

$$g\sigma : x \mapsto xg \, (x \in G).$$

Straightforward verification establishes that σ is a faithful permutation representation of G. $\qquad\square$

This representation is called the *right regular representation*.

4.2 Frobenius' Permutation Representation

Let G be a group and A a subgroup of G. A *complete set of representatives* of the right cosets Ag of A in G is a subset $Y \subseteq G$ satisfying:

K.A. Bencsáth et al., *Lectures on Finitely Generated Solvable Groups*,
SpringerBriefs in Mathematics, DOI 10.1007/978-1-4614-5450-2_4,
© Katalin A. Bencsath, Marianna C. Bonanome, Margaret H. Dean, Marcos Zyman 2013

1. For every $g \in G$, there exists $y \in Y$ such that $Ag = Ay$.
2. $Ay_1 = Ay_2$ for $y_1, y_2 \in Y$ implies that $y_1 = y_2$.

The element of Y coming from the coset Ag is denoted by $\overline{g}$ and it is called the representative of the coset Ag. If $1 \in Y$, Y is called a *right transversal* of A in G.

Suppose g, z are elements of G. Put $\delta(z, g) = zg(\overline{zg})^{-1}$. Intuitively, δ registers how the representative $\overline{zg}$ deviates from the actual product zg.

Lemma 4.1.

1. $Ag = A\overline{g} \ (g \in G)$.
2. $\delta(z, g) \in A \ (z, g \in G)$.
3. $\overline{\overline{g_1}g_2} = \overline{g_1 g_2} \ (g_1, g_2 \in G)$.
4. $g = \delta(g, 1)\overline{g} \ (g \in G)$.

Lemma 4.1 can be verified directly from the definitions.

Theorem 4.2 (Frobenius). *Let G be a group, A a subgroup of G, and Y a complete set of representatives of right cosets of A in G. The following hold:*

1. *Each $g \in G$ gives rise to a permutation $\hat{g} : Y \to Y$, where*

$$y\hat{g} = \overline{yg}.$$

2. *The map $\varphi : G \to S_Y$ given by*

$$g\varphi = \hat{g}$$

   is a homomorphism from G into the permutation group of Y.
3. *The kernel of φ is the largest normal subgroup of G contained in A, called the* **core** *of A in G.*
4. *If A is normal in G, φ induces a faithful permutation representation of G/A in S_Y.*
5. *If G is simple (i.e., G has no nontrivial normal subgroup), then φ is faithful.*

Proof. 1. First, we show that $\hat{g} \, (g \in G)$ is one-to-one. Suppose that for elements $y_1, y_2 \in Y$, $\overline{y_1 g} = \overline{y_2 g}$. Then

$$Ay_1 g = Ay_2 g,$$

so $Ay_1 = Ay_2$. Since Y contains exactly one representative from each coset, it follows that $y_1 = y_2$. Thus, $\hat{g}$ is one-to-one.

To prove that $\hat{g}$ is onto, let $y \in Y$, and observe that by Lemma 4.1

$$\left(\overline{yg^{-1}}\right)\hat{g} = \overline{\overline{yg^{-1}}g} = \overline{yg^{-1}g} = \overline{y} = y.$$

Thus, $\hat{g}$ is onto as well, so $\hat{g}$ is a permutation of Y.
2. To show that φ is a homomorphism, it suffices to check that for all pairs of elements g_1, g_2 of G, $\hat{g_1}\hat{g_2} = \widehat{g_1 g_2}$ in S_Y.

Direct computation (with $y \in Y$) while keeping the definition of φ and Lemma 4.1 in mind confirms:

$$y\hat{g_1}\hat{g_2} = (y\hat{g_1})\hat{g_2} = \overline{\overline{yg_1}g_2} = \overline{\overline{yg_1 g_2}} = \overline{yg_1 g_2} = y\widehat{g_1 g_2}.$$

3. Let N denote the core of A. Suppose that $g \in ker\varphi$. $\hat{g} = id_{S_Y}$ is equivalent to $Ayg = Ay$ for all $y \in Y$. If $y_0 \in Y$ is the representative of A, this means $Ay_0 g = Ay_0 = A$, so $y_0 g \in A$. Since $y_0 \in A$, $g \in A$ and $ker\varphi \subseteq A$ follow. As $ker\varphi$ is one of the normal subgroups of G contained in A, $ker\varphi \subseteq N$.

As the condition $ygy^{-1} \in A$ for all $y \in Y$, which is equivalent to $g \in ker\varphi$, is satisfied by all elements g belonging to any normal subgroup of G contained in A, we also have the reverse containment $N \subseteq ker\varphi$. Therefore, $ker\varphi$ is the core of A, indeed.

Assertions (4) and (5) are immediate. $\square$

4.3 A Theorem of Krasner and Kaloujnine

Recall that the construction of the unrestricted wreath product (Sect. 3.2.2) started with a group A and an action of a group T on some set Y, that induced an action φ on the unrestricted direct product A^Y:

$$\alpha\, t\varphi(y) = \alpha\left(yt^{-1}\right) \quad (y \in Y,\ \alpha \in A^Y, t \in T).$$

Let X be a new set, on which the group A acts. The Cartesian product $X \times Y$ can be viewed as the disjoint union of "blocks" $X \times \{y\}$ ($y \in Y$) indexed by the elements of Y, each a copy of X. For each $y \in Y$, there is an action of A^Y on the block $X \times \{y\}$ given by

$$(x,y)\alpha = (x\,\alpha(y),y),$$

which permutes the elements within $X \times \{y\}$ while keeping the block in place.

For $t \in T$, there is an action

$$(x,y)t = (x,yt)$$

on $X \times Y$ induced by the action of T on Y which moves any given block with index y to the X-block whose index is the image of y under the permutation of Y caused by $t \in T$.

Proposition 4.1. *The above described pair of actions of A^Y and T on $X \times Y$ induce an action of $\overline{W} = A^Y \rtimes_\varphi T$ on $X \times Y$ defined as follows:*

$$(x,y)(t\alpha) = (x\,\alpha(yt),yt) \quad (\alpha \in A^Y,\ t \in T).$$

Proof. Since for each $w = t\alpha$ the mapping proposed above is clearly a permutation of the set $X \times Y$, it suffices to show that

$$(x,y)(w_1 w_2) = ((x,y)w_1)\,w_2 \quad \left(w_1, w_2 \in \overline{W},\ (x,y) \in X \times Y\right). \tag{4.1}$$

Let $w_1 = t_1\alpha_1$ and $w_2 = t_2\alpha_2$ expressed in their unique semi-direct normal forms in $\overline{W}$. Then the unique semi-direct normal form for their product in $\overline{W}$ is

$$w_1 w_2 = t_1 t_2 \alpha_1^{t_2} \alpha_2.$$

Therefore, the left hand side of (4.1) is

$$(x,y)(w_1 w_2) = \left(x\left(\alpha_1^{t_2}\alpha_2(yt_1 t_2)\right), yt_1 t_2\right)$$
$$= \left(x\left(\alpha_1^{t_2}(yt_1 t_2)\alpha_2(yt_1 t_2)\right), yt_1 t_2\right) \in X \times Y,$$

while the right hand side of (4.1) is

$$((x,y)w_1)w_2 = ((x,y)t_1\alpha_1)t_2\alpha_2 = (x\alpha_1(yt_1), yt_1)t_2\alpha_2$$
$$= (x[\alpha_1(yt_1)\alpha_2(yt_1 t_2)], yt_1 t_2).$$

By virtue of the definition of multiplication in A^Y, the equality

$$\alpha_1^{t_2}(yt_1 t_2)\alpha_2(yt_1 t_2) = \alpha_1(yt_1)\alpha_2(yt_1 t_2)$$

follows from the action of T on A^Y:

$$\alpha_1^{t_2}(yt_1 t_2) = \alpha_1(yt_1 t_2 t_2^{-1}) = \alpha_1(yt_1).$$

$$\square$$

Proposition 4.2. *If the actions of T on Y and A on X are both faithful, then $\overline{W}$ acts faithfully on $X \times Y$.*

Proof. Assume that the actions of T on Y and A on X are faithful. Let $t\alpha$ be an element of $\overline{W}$. Since $\overline{W}$ acts on $X \times Y$, we can identify $t\alpha$ with a permutation of $X \times Y$. We need to prove that if $t\alpha$ is the identity permutation, then $t\alpha = 1$ in $\overline{W}$. But if $t\alpha$ is the identity permutation, then for all (x,y) in $X \times Y$ we have

$$(x,y)(t\alpha) = (x\alpha(yt), yt) = (x,y).$$

This implies that $yt = y$ and $x\alpha(yt) = x$. Since the actions of A on X and T on Y are faithful, we conclude that $t = 1$ in T, and $\alpha(y) = 1$ in A (for every $y \in Y$). That is to say, α is the identity element $(1,1,1,\dots)$ of A^Y. $\square$

We are now in position to prove the theorem of Krasner and Kaloujnine. This remarkable theorem states that any non-simple group can be viewed as a subgroup of an unrestricted wreath product. The proof is based on recasting Cayley's representation in a manner that makes use of a given Frobenius representation.

Theorem 4.3. *Let*

$$1 \to A \to G \to H \to 1$$

be a short exact sequence of groups. Then G can be embedded in an unrestricted wreath product

$$\overline{W} = A \,\bar{\wr}\, H.$$

Proof. In order to prove this result, we begin with a more general setup. Let $A \leq G$, not necessarily normal, and let Y be a complete set of representatives of the right cosets of A in G. Every element of G can be uniquely expressed as a product ay where $a \in A$ and $y \in Y$. Moreover, using Frobenius' machinery and denoting $\delta(y,g)$ by $\delta_g(y)$ (see Sect. 4.2), we have that for each $g \in G$

$$(ay)g = a(yg) = a\delta_g(y)\overline{yg}.$$

Note that for each $g \in G$, we may regard δ_g as an element of A^Y [see Lemma 4.1 (2)]. Since G and $A \times Y$ are set-isomorphic and since $\delta_g(y) \in A$, Cayley's representation of G in S_G (Theorem 4.1) induces a faithful representation of G in $S_{A \times Y}$. This gives rise to a faithful action of G on $A \times Y$ as follows:

$$(a,y)g = (a\delta_g(y), \overline{yg}) \in A \times Y \quad ((a,y) \in A \times Y,\ g \in G).$$

Write

$$\rho_1 : G \hookrightarrow S_{A \times Y}$$

for the faithful representation associated to this action.

On the other hand, Frobenius' representation (Theorem 4.2) tells us that G has an action on Y

$$yg = \overline{yg} \in Y$$

and that the kernel K of that action is exactly the core of A. So, writing H for G/K, the action factors through H to give a faithful action of H on Y in the obvious way:

Form the unrestricted wreath product

$$\overline{W} = A \,\bar{\wr}\, H.$$

A acts on itself by right multiplication. Thus, we have the following faithful actions on $A \times Y$:

1. For $\alpha \in A^Y$, $(a,y)\alpha = (a\alpha(y), y)$.
2. For $h \in H$, $(a,y)h = (a, \overline{yh})$.

Hence, by Proposition 4.1 and 4.2, $\overline{W}$ acts faithfully on $A \times Y$. Let

$$\rho_2 : \overline{W} \hookrightarrow S_{A \times Y}$$

be the corresponding monomorphism.

In summary, we have two faithful permutation representations:

- $\rho_1 : G \hookrightarrow S_{A \times Y}$
- $\rho_2 : \overline{W} \hookrightarrow S_{A \times Y}$

The remainder of the proof lies in the argument that $Im(\rho_1) \leq Im(\rho_2)$, which permits the conclusion that there exists an embedding

$$\rho : G \hookrightarrow \overline{W}.$$

Let $g \in G$ and identify it with its image $g\rho_1$ in $S_{A \times Y}$. Then g is the permutation of $A \times Y$ given by

$$(a,y)g = (a\delta_g(y), \overline{yg}) \quad ((a,y) \in A \times Y).$$

Consider $\delta_g \in A^Y$ and (with a slight abuse of notation) $\bar{g} \in H = G/K$. Then $w = (\bar{g}, \delta_g) \in \overline{W}$. If we identify w with its image $w\rho_2$ in $A \times Y$ we obtain:

$$(a,y)w = (a,y)(\bar{g}, \delta_g) = (a\delta_g(y), \overline{yg}) = (a,y)g.$$

Hence, we have $G \cong Im(\rho_1) \leq Im(\rho_2) \cong \overline{W}$, which provides the desired embedding

$$\rho : G \hookrightarrow \overline{W}.$$

Finally, choose A normal in G; then $H = G/A$ and the theorem follows. $\square$

Remark 4.1. If A is of finite index in G, then $\overline{W} = W$, and G can be embedded in a restricted wreath product $W = A \wr H$ (see Sect. 3.2.1).

Chapter 5
The Bieri–Strebel Theorems

At last, we are ready to begin our investigation of finitely generated solvable groups: we are now prepared to prove two theorems of Bieri and Strebel. These provide an important step in the understanding of finitely presented solvable groups. The results of this chapter are developed with no particular assumptions on the (finite) solvability lengths of the (finitely generated) solvable groups investigated here.

Recall that an HNN-extension $G = \langle B, t;\ t^{-1}Ht = K \rangle$ is termed ascending if one of the associated subgroups coincides with the base group B (see Sect. 3.3.1). If both H and K coincide with B, then t acts as an automorphism of B (which is not too interesting); we will be more concerned with the situation in which exactly one of the associated subgroups coincides with B.

We begin with an easy but potent lemma.

Lemma 5.1. *If $G = \langle B, t;\ t^{-1}Ht = K \rangle$ is an HNN-extension which is not ascending, then G contains a free subgroup on two generators.*

Proof. We need to find two elements in G that generate a non-abelian free subgroup of G. Choose b_1 and $b_2 \in B$ such that $b_2 \notin H, b_1 \notin K$. Since $H \neq B, K \neq B$, that is possible to do. b_1 and b_2 need not be distinct.

Consider

$$u = b_1 t^{-2} b_2 t^2, \quad v = t b_1 t^{-2} b_2 t.$$

We claim that u and v freely generate a free subgroup. First, we check that u, v each have infinite order. Suppose that $u^2 = b_1 t^{-2} b_2 t^2 b_1 t^{-2} b_2 t^2 = 1$. The only candidates for pinches within u^2 are $t^{-1} b_2 t$ or $t b_1 t^{-1}$. But neither of these are actually pinches since $b_2 \notin H$ and $b_1 \notin K$. This generalizes easily; for any nonzero integer n, $u^n \neq 1$. Similar arguments show v also has infinite order, and that $u^{n_1} v^{m_1} \cdots u^{n_k} v^{m_k} \neq 1$ unless $n_1 = \cdots = n_k = m_1 = \cdots = m_k = 0$. Hence, $gp(u,v)$ is a free subgroup of rank 2. Of course, there are many other choices possible for u and v. $\qquad\square$

Theorem 5.1 is the first of the Bieri–Strebel theorems; its relevance to solvability is twofold: first, it offers a condition and a construction for determining that certain groups are not solvable; the second impact will become apparent in Theorem 5.2.

K.A. Bencsáth et al., *Lectures on Finitely Generated Solvable Groups*,   25
SpringerBriefs in Mathematics, DOI 10.1007/978-1-4614-5450-2_5,
© Katalin A. Bencsath, Marianna C. Bonanome, Margaret H. Dean, Marcos Zyman 2013

Theorem 5.1 (Bieri and Strebel). *Let N be a normal subgroup of a finitely presented group G. If G/N is infinite cyclic, then G is an ascending HNN-extension of a finitely generated group or contains a free subgroup of rank 2.*

Proof. We will show that any finitely presented group that has the infinite cyclic group as its homomorphic image is an HNN-extension of a finitely generated group.

Since G/N is infinite cyclic, we can choose $t \in G$ of infinite order, such that $G/N = gp(tN)$. In particular, this implies that $G = gp(t,N)$. We need to concoct a base group, with stable letter t. Note that $G = gp(t,a_1,\ldots)$, where (setwise) $\{a_1,\ldots\} = N$. Since G is finitely generated, G is generated by finitely many of these elements; say, $G = gp(t,a_1,\ldots,a_k)$. In addition, G is finitely presented, so by Lemma 2.2 we can present it on these generators:

$$G = \langle t,a_1,\ldots,a_k ; r_1,r_2,\ldots,r_l \rangle.$$

Each r_j is a word in t and $a_1,\ldots,a_k$. Observe that because G/N is infinite cyclic, generated by tN, t must occur with exponent sum 0 in all of the r_j.

With the insertion of appropriate powers of t, proceeding from right to left, put $a_{p,m} = t^{-m}a_p t^m$ $(1 \le p \le k, m \in \mathbb{Z})$. The r_j $(j = 1,\ldots,l)$ can be rewritten so each becomes a word in the $a_{p,m}$'s, as the following example illustrates:

$$\begin{aligned}
r &= t^2 a_1^3 a_2^{-1} t a_3^{-1} t a_3^2 a_1 t^{-4} a_2 \\
&= t^2 a_1^3 a_2^{-1} t a_3^{-1} t t^{-4} t^4 a_3^2 a_1 t^{-4} a_2 \\
&= t^2 a_1^3 a_2^{-1} t^{-2} t^2 t a_3^{-1} t^{-3} a_{3,-4}^2 a_{1,-4} a_{2,0} \\
&= a_{1,-2}^3 a_{2,-2}^{-1} a_{3,-3}^{-1} a_{3,-4}^2 a_{1,-4} a_{2,0}.
\end{aligned}$$

Now we use Tietze transformation **T1** (see Theorem 2.1) to "enlarge" the presentation for G, as follows:

$$\begin{aligned}
G = \langle t,a_1,\ldots,a_k,a_{p,m} \quad &(1 \le p \le k, \ m \in \mathbb{Z}); r_1,\ldots,r_l, \\
a_{p,m} = t^{-m}a_p t^m \quad &(1 \le p \le k, \ m \in \mathbb{Z}) \rangle.
\end{aligned}$$

Each relation $a_{p,m} = t^{-m}a_p t^m$ $(1 \le p \le k, m \in \mathbb{Z})$ is a consequence of $a_{p,0} = a_p$ and $a_{p,m} = t^{-1}a_{p,m-1} t$ $(1 \le p \le k, m \in \mathbb{Z})$. Thus, we can apply **T2** to obtain the alternate presentation

$$\begin{aligned}
G = \langle t,a_1,\ldots,a_k,a_{p,m} \quad &(1 \le p \le k, m \in \mathbb{Z}); r_1,\ldots,r_l, \\
a_{p,0} = a_p, a_{p,m} = t^{-1}a_{p,m-1} t \quad &(1 \le p \le k, m \in \mathbb{Z}) \rangle.
\end{aligned}$$

As it stands, this is an infinite presentation. But since there are finitely many r_j, only finitely many values of m are actually needed. Hence, there are fixed integers $q \le 0$

and s such that all the r_j can be written as words in the $a_{p,m}$ $(q \leq m \leq s)$. Use **T1'** to "shrink" the presentation for G:

$$G = \langle t, a_1, \ldots, a_k, a_{p,m} \quad (1 \leq p \leq k, q \leq m \leq s); r_1, \ldots, r_l,$$

$$a_{p,0} = a_p, \ a_{p,m} = t^{-1} a_{p,m-1} t \quad (1 \leq p \leq k, \ q < m \leq s) \rangle.$$

Put r for $-q$ and note that r is nonnegative; further, denote the positive integer $s + r$ by δ. Using **T1** and **T2**, we can add $a_{p,m+r}$ $(1 \leq p \leq k, q \leq m \leq s)$ to the list of generators and $a_{p,m} = t^{-1} a_{p,m-1} t$ $(1 \leq p \leq k, q + r < m \leq \delta)$ to the list of relators; we can then use **T1'** and **T2'** to remove generators and relators so that G can be presented thus:

$$G = \langle t, a_1, \ldots, a_k, a_{p,m} \quad (1 \leq p \leq k, \ 0 \leq m \leq \delta); r_1, \ldots, r_l,$$

$$a_{p,0} = a_p, a_{p,m} = t^{-1} a_{p,m-1} t \quad (1 \leq p \leq k, \ 1 \leq m \leq \delta) \rangle.$$

Using **T1'**, we can now throw out the old $a_1, \ldots, a_k$ and find that

$$G = \langle t, a_{p,m} \quad (1 \leq p \leq k, 0 \leq m \leq \delta); r_1, \ldots, r_l,$$

$$a_{p,m} = t^{-1} a_{p,m-1} t \ (1 \leq p \leq k, \ 1 \leq m \leq \delta) \rangle.$$

To see that this is an *HNN*-extension of a finitely generated group, set

$$B = gp(a_{p,m} \mid 1 \leq p \leq k, \ 0 \leq m \leq \delta),$$

$$H = gp(a_{p,y} \mid 1 \leq p \leq k, \ 0 \leq y \leq \delta - 1)$$

and

$$K = gp(a_{p,x} \mid 1 \leq p \leq k, \ 1 \leq x \leq \delta).$$

Thus,

$$G = \langle B, t; H^t = K \rangle.$$

If this *HNN*-extension is not ascending, then G contains a free subgroup of rank 2 by Lemma 5.1. □

Remark 5.1. On the face of it, given the generators for H and K, neither one appears to equal B. However, in truth we know very little about the interactions among the $a_{p,m}$. Either H or K or both could in fact equal B.

There is a partial converse to Theorem 5.1, which is easy to prove.

Proposition 5.1. *If G is an ascending HNN-extension, then G has a normal subgroup N such that G/N is infinite cyclic.*

Proof. Suppose that $K = B$. Then $t^{-1}Ht = B$, or $H = tBt^{-1}$. Now, $H \leq B$, so

$$B = t^{-1}Ht \leq t^{-1}Bt.$$

Put $B_i = t^{-i}Bt^i$. Then we find that $B = B_0 \leq B_1 \leq \dots$. Hence,

$$gp_G(B) = \bigcup_{i=0}^{\infty} B_i$$

and $G/gp_G(B) \cong \mathbb{Z} \cong \langle t \rangle$. $\qquad\qquad\qquad\qquad\qquad\qquad\qquad\qquad\qquad\square$

Remark 5.2. Example 3.9 in Sect. 3.3.3 is a useful illustration of how G might look.

The proof of Theorem 5.2 relies on the following two lemmas. Lemma 5.2 is a general result used frequently in combinatorial group theory.

Lemma 5.2. *If H is a subgroup of finite index in a finitely generated group G, then H is finitely generated.*

Proof. Let Y be a right transversal of H in G. Denote the representative of $a \in G$ by $\bar{a}$. As in Sect. 4.2, set $\delta(a,b) = ab(\overline{ab})^{-1}$. Then $\delta(a,b) \in H$.

Let X be a set of generators for G. Recall (Sect. 4.3) that G acts on $H \times Y$: $(h,y)g = (h\delta(y,g),\overline{yg})$.

If $g \in G$, then $g = a_1 a_2 \cdots a_n$ $(a_i \in X \cup X^{-1})$. Compute the action of g on $(1,1)$ in the event that $g \in H$:

$$(1,1)g = (\delta(1,g),\overline{1g}) = (g(\bar{g})^{-1},1) = (g,1).$$

Thus,

$$(1,1)a_1 a_2 \cdots a_n = (\delta(1,a_1),\bar{a}_1)a_2 \cdots a_n = \cdots = (g,1).$$

The first coordinate is a product of elements from Y and elements from X, and this product equals g. Since Y and X are both finite, we have found a finite set of generators for H. $\qquad\qquad\qquad\qquad\qquad\qquad\qquad\qquad\qquad\square$

Lemma 5.3. *Let G be a finitely generated solvable group. Then G is either finite or contains a subgroup H of finite index which in turn contains a normal subgroup N with H/N infinite cyclic.*

Proof. Suppose G is not finite. Consider the first term $[G,G]$ of the derived series for G. Since $G/[G,G]$ is a finitely generated abelian group, it is the direct product of a finite number of cyclic groups. Either at least one of those cyclic groups is infinite cyclic, or $[G,G]$ is of finite index in G.

If $[G,G]$ is of infinite index, $G/[G,G]$ is infinite and

$$G/[G,G] = M/[G,G] \times C_1/[G,G] \times \cdots \times C_n/[G,G],$$

where $M/[G,G]$ is infinite cyclic and each of the other factors is cyclic (finite or infinite). Let N be the normal closure of $C_1,\dots,C_n$ in G. Then $G/N \cong \mathbb{Z}$ and G itself is the required subgroup H of finite index in G.

If $[G,G]$ is of finite index, then $G/[G,G]$ is finite. We apply the same argument to $[G,G]$, which we know is finitely generated by Lemma 5.2. Take G'/G'', a finitely generated abelian group, and so on. Iterate this process down the derived series until we obtain the groups H and N that we seek. The procedure will end before we exhaust all the terms of the derived series; or else, G would be finite, contrary to our assumption.   □

Theorem 5.2 (Bieri and Strebel). *Let G be an infinite, finitely presented solvable group. Then G contains a subgroup of finite index which is an ascending HNN-extension of a finitely generated solvable group.*

Proof. By the proof of Lemma 5.3, for some nonnegative integer n, the nth derived group G^n is of finite index in G. In addition, there is a normal subgroup N of G^n such that H/N is infinite cyclic. G^n is finitely generated, by Lemma 5.2.

Now, for any group G, every member of the derived series of G is fully invariant; that is, if $G = F/R$ for some free group F, then $G^n = F^n R/R$. It follows that if G is finitely presented, G^n is finitely presented, and by Theorem 5.1, Propositions 1.1, and 1.2, G^n is an ascending HNN-extension of a finitely generated solvable group B.   □

Remark 5.3. Theorem 5.2 shows that the structure of an infinite finitely presented solvable group G is enormously restricted. It guarantees that G is always virtually an ascending HNN-extension of a finitely generated group.

Next, we revisit Example 3.3, and use Theorem 5.1, to show that M is not finitely presented.

Example 5.1. Let $\Lambda = \mathbb{Z}[1/6]$, the ring of integers with $\frac{1}{6}$ adjoined. Observe that Λ is a subring of $\mathbb{Q}$, and note that $\frac{3}{2}$ is a unit of Λ. Let $A = \Lambda^+ = \left\{ \frac{l}{6^m} \mid l, m \in \mathbb{Z} \right\}$ be the additive group of Λ. As in the case of the additive group of rationals, A is not finitely generated but every finitely generated subgroup of it is infinite cyclic.

Following the construction used in Sect. 3.1.3, let $\langle t \rangle$ be an infinite cyclic group generated by t, and consider the homomorphism $\varphi : \langle t \rangle \to Aut\,A$ determined by

$$t\varphi : a \mapsto \frac{3}{2}a \ (a \in A).$$

Form the semi-direct product

$$M = A \rtimes_\varphi \langle t \rangle.$$

A typical element of M can be written as $\left(t^i, \frac{l}{6^m} \right)$, and multiplication in M is given by:

$$\left(t^i, \frac{l}{6^m} \right) \left(t^j, \frac{l'}{6^{m'}} \right) = \left(t^{i+j}, \left(\frac{3}{2} \right)^j \left(\frac{l}{6^m} \right) + \frac{l'}{6^{m'}} \right).$$

For ease of notation, henceforth we will treat M as an internal semi-direct product. If c denotes the integer 1, it is easy to prove that $M = gp(t,c)$. In addition, by putting $c_i = t^{-i}ct^i$ $(i \in \mathbb{Z})$, it follows that

$$A = gp(\ldots, c_{-2}, c_{-1}, c_0, c_1, c_2, \ldots).$$

From the short exact sequence

$$0 \longrightarrow A \longrightarrow M \longrightarrow \langle t \rangle \longrightarrow 1,$$

we see that M is an extension of the abelian subgroup A by the infinite cyclic group $\langle t \rangle$; hence, metabelian.

Suppose that M is finitely presented. By Proposition 1.1, subgroups of solvable groups are solvable, and by Proposition 1.2, a metabelian group cannot have a free subgroup of rank 2. Further, $M/A \cong \langle t \rangle$, which is infinite cyclic. By Theorem 5.1, M must be an ascending HNN-extension, with stable letter t and base group B, a finitely generated subgroup of A. But all finitely generated subgroups of A are infinite cyclic. So $B = gp(d)$ for some $d \in A$ of infinite order. In an ascending HNN-extension, one of the associated subgroups must

$$t^{-1}Bt \leq B \quad \text{or else} \quad tBt^{-1} \leq B.$$

Suppose $t^{-1}Bt \leq B$. Then $t^{-1}dt = \frac{3}{2}d \in B = gp(d)$. But this implies that $3/2 \in \mathbb{Z}$, which is not true. The other case is similarly impossible. Hence, M is not finitely presented.

Remark 5.4. If we express the group operation in A as multiplication, as we did in Sect. 3.1.3, conjugation by t yields $t^{-1}c^2t = c^3$.

Remark 5.5. This example is one of an infinite family of examples, wherein we start with the ring $\Lambda = \mathbb{Z}\left[\frac{1}{mn}\right]$ with $m/n \notin \mathbb{Z}$.

Chapter 6
Finitely Generated Metabelian Groups

In the remaining two chapters, we turn our attention exclusively to finitely generated metabelian groups. First we introduce some terminology and auxiliary results.

Lemma 6.1. *Suppose that G is a group generated by the set X. Then*

$$[G,G] = gp_G\left([x,y] \mid x,y \in X\right).$$

Proof. Let $H = gp_G\left([x,y] \mid x,y \in X\right)$. Since $[G,G]$ is normal in G, H is a subgroup of $[G,G]$. On the other hand, it is evident that G/H is abelian, so if we consider any generator $[u,v]$ of $[G,G]$, we see that in G/H,

$$[u,v]H = H.$$

Therefore $[u,v] \in H$, and $[G,G] \leq H$. Thus, $H = [G,G]$. □

An immediate corollary is the following:

Corollary 6.1. *If G is finitely generated, $[G,G]$ is finitely generated as a normal subgroup.*

Lemma 6.2. *Let G be a group and A an abelian, normal subgroup of G. Then the quotient group $Q = G/A$ acts on A.*

Proof. To show that Q acts on A, we exhibit a homomorphism φ from Q to $Aut(A)$. For every $gA \in Q$, we define

$$(gA)\varphi = \{a \mapsto a^g \ (a \in A)\}.$$

Then φ is well-defined; for if $gA = hA$, then $g = hb$ for some $b \in A$. Since A is abelian and normal, for each $a \in A$ we have:

$$a^g = g^{-1}ag = b^{-1}h^{-1}ahb = h^{-1}ah = a^h.$$

K.A. Bencsáth et al., *Lectures on Finitely Generated Solvable Groups,*
SpringerBriefs in Mathematics, DOI 10.1007/978-1-4614-5450-2_6,
© Katalin A. Bencsath, Marianna C. Bonanome, Margaret H. Dean, Marcos Zyman 2013

It is also easy to check that $(ghA)\varphi = (gA)\varphi(hA)\varphi$, hence φ is a homomorphism, as required.   □

A few facts about rings and modules are needed here, before we proceed.

Recall that an *endomorphism* of a group is a homomorphism from the group to itself. If M is an abelian group, then the set of all endomorphisms of M is a unital ring, denoted by $End(M)$.

Unital ring homomorphisms map the multiplicative identities of the rings to each other. Unless explicitly stated otherwise, the ring homomorphisms encountered here will be unital.

Suppose now that R is a ring with unity. An abelian group M is called a *(right) R-module* if it comes equipped with a unital homomorphism ψ from the ring R to $End(M)$. In such a case, we use the notation $m(r\psi) = mr$ and say that R acts on M from the right.

We also recall that if Q is any group, then the set of all functions from Q to $\mathbb{Z}$ with finite support,

$$\mathbb{Z}^{(Q)} = \{f : Q \to \mathbb{Z} \mid f(q) = 0 \text{ for all but finitely many } q \in Q\},$$

turns into a ring upon defining

$$f + f' : q \mapsto f(q) + f'(q) \quad (q \in Q),$$

and

$$ff' : q \mapsto \sum_{q_1 q_2 = q} f(q_1) f'(q_2) \quad (q, q_1, q_2 \in Q).$$

We denote this (not necessarily commutative) ring by $\mathbb{Z}Q$ and refer to it as the integral group ring of Q. Additively, it is a free abelian group on the set underlying the group Q, and it contains (a copy of) $\mathbb{Z}$ as the infinite cyclic group generated by 1 (the function that sends the identity of Q to 1 and all other elements of Q to 0).

Lemma 6.3. *Let A be an abelian, normal subgroup of a finitely generated group G, and suppose that A is finitely generated as a normal subgroup in G. Then the action of $Q = G/A$ on A described in Lemma 6.2 turns A into a finitely generated $\mathbb{Z}Q$-module.*

Proof. According to Lemma 6.2, $Q = G/A$ acts on the abelian group A as follows:

$$a(gA) = g^{-1}ag \ (a \in A).$$

This induces an action of the group ring $\mathbb{Z}Q$ on A, and thus a $\mathbb{Z}Q$-module structure on A. Since A is finitely generated as a normal subgroup of G, every given $a \in A$ can be expressed in terms of finitely many G-conjugates of a fixed finite set of, say, k elements $g_1, \ldots, g_k$ of A. Therefore, writing A additively, we have

$$a \left(\sum_{i=1}^{k} m_i g_i A \right) = \sum_{i=1}^{k} m_i \left(g_i^{-1} a g_i \right) \quad (g_i \in G, \, m_i \in \mathbb{Z}).$$

Thus, A can be viewed as a finitely generated $\mathbb{Z}Q$-module. $\quad\square$

In the remainder of these notes, we are going to make free use of the following fact from commutative algebra (see [1], Chap. 7):

Since $\mathbb{Z}Q$ is Noetherian, every finitely generated submodule of a finitely generated $\mathbb{Z}Q$-module M is finitely generated.

Lemma 6.4. *Let N be a normal subgroup of a finitely generated metabelian group. Then N is finitely generated as a normal subgroup.*

Proof. Let G be a finitely generated metabelian group. Then $G' = [G, G]$ is abelian and normal. By Corollary 6.1, G' is finitely generated as a normal subgroup, and by Lemma 6.3, G' becomes a finitely generated $\mathbb{Z}(G/G')$-module. Since G is finitely generated, so is the abelian group G/G'. Thus, we can regard G' as a finitely generated module over a polynomial ring in finitely many variables with coefficients from $\mathbb{Z}$.

For any normal subgroup N of G, the subgroup $N \cap G'$ of G' is in fact a submodule of G', over said polynomial ring in finitely many variables with coefficients from $\mathbb{Z}$. Thus, $N \cap G'$ is finitely generated as a submodule, say by $\{a_1, \ldots, a_p\}$.

In addition, since G/G' is a finitely generated abelian group, so is its subgroup NG'/G'. Thus, there exist elements $\{b_1, \ldots b_q\}$ in N whose images generate NG'/G'. It follows that N is the normal closure of a finite set; indeed,

$$N = gp_G(a_1, \ldots, a_p, b_1, \ldots, b_q).$$

$\quad\square$

Corollary 6.2. *Every finitely generated metabelian group is finitely presented as a metabelian group.*

Proof. Let G be a finitely generated group in the class of all metabelian groups. In this class, we can write a presentation

$$G = \langle\langle X; R \rangle\rangle.$$

It is standard to use "$\langle\langle \; \rangle\rangle$" to mean that $G \cong F(X)/gp_F(R)$, where $F(X)$ is the free metabelian group generated by a finite set X; and $gp_F(R)$ is the normal closure of R in $F(X)$. $gp_F(R)$ is a normal subgroup of the finitely generated metabelian group $F(X)$, so it is the normal closure of a finite set by Lemma 6.4. $\quad\square$

Even finitely generated free metabelian groups fail to be finitely presented; infinitely many relators are needed to encode the metabelian rule that commutators commute. Thus, Corollary 6.2 is only true within the class of metabelian groups.

A group G is said to satisfy *Max-n* if every properly ascending chain of normal subgroups of G is finite.

Theorems 6.1 and 6.2 are due to P. Hall:

Theorem 6.1. *Finitely generated metabelian groups satisfy Max-n.*

Proof. Let G be a finitely generated metabelian group, and consider a properly ascending chain of normal subgroups of G

$$N_1 < N_2 < \cdots .$$

Let

$$N = \cup N_i,$$

which is a normal subgroup of G. By Lemma 6.4, N must be the normal closure of a finite set. Hence, there exists an index j and finitely many elements $x_1, \ldots, x_m \in N_j$ such that

$$N = gp_G(x_1, \ldots, x_m).$$

N_j is normal in G; thus,

$$N = N_j,$$

so the chain is finite indeed. $\square$

Theorem 6.2. *There are only countably many isomorphism classes of finitely generated metabelian groups.*

Proof. It suffices to show that for a given $n < \infty$, there are only countably many n-generator metabelian groups. Let G be an n-generator metabelian group. Then there exists a normal subgroup N of F, the free metabelian group of rank n, such that

$$G \cong F/N.$$

By Lemma 6.4, N is the normal closure of a finite set in F. Therefore, the number of n-generator metabelian groups is bounded by the number of finite subsets of the countable group F, which is countable. $\square$

Example 6.1. Consider the group G from Example 3.3. G is an extension of an abelian group by an abelian group; hence, metabelian. As we shall see in Chap. 7, G is not finitely presented. However, Corollary 6.2 says that G is finitely presented as a metabelian group. A metabelian presentation is

$$G = \langle\langle a,t;\ [a,t]^a = [a,t],\ [a,t] = a^2 a^{-t}\rangle\rangle .$$

If we are given that G' is abelian, we only need to view G' as a finitely generated $\mathbb{Z}(G/G')$-module, and put in the finitely many relations required to describe this module, as well as the relations that yield a finite presentation of G/G'. The reader is encouraged to consult [4, 5] for a more detailed treatment of the procedure.

Chapter 7
An Embedding Theorem for Finitely Generated Metabelian Groups

Not all finitely generated metabelian groups are finitely presented (see Example 7.1). However, Baumslag's embedding theorem [3] establishes that, in fact, every finitely generated metabelian group can be embedded in a finitely presented metabelian group (see Theorem 7.1). This is a remarkable refinement on Higman's result from 1961 [11] concerning embeddings of finitely generated groups into finitely presented groups.

A natural question arises regarding whether we can turn Theorem 5.2 around to create a finitely presented solvable group by making an ascending HNN-extension from a finitely generated solvable base group B.

This embedding theorem links the theories of finitely generated metabelian groups and subgroups of finitely presented metabelian groups.

Remark 7.1. Theorem 7.1 marks another sharp difference between metabelian groups and solvable groups of higher solvability length. So far, the different universe has a known boundary at the class of center-by-metabelian groups. The existence of continuously many non-isomorphic finitely generated center-by-metabelian groups (see Theorem 1.1) combined with the fact that the number of non-isomorphic finitely presented groups is countable, leads to the conclusion [2] that there can be no such embedding theorem for center-by-metabelian groups.

We start with an example of a finitely generated metabelian group which is not finitely presented, but forming an HNN-extension embeds it into a finitely presented metabelian group. This will give us an indication of how the proof of Theorem 7.1 works.

Example 7.1. Let B be the free abelian group on $\{\ldots, a_{-1}, a_0, a_1, \ldots\}$ and $\langle t \rangle$ an infinite cyclic group generated by t. Let $W = B \rtimes \langle t \rangle$ where t acts on B by translation; i.e., $t^{-1} a_i t = a_{i+1}$ $(i \in \mathbb{Z})$ (see Examples 3.4 and 3.5). Recall that $W = \langle a \rangle \wr \langle t \rangle$. W is an abelian-by-abelian group; hence, metabelian.

K.A. Bencsáth et al., *Lectures on Finitely Generated Solvable Groups*,
SpringerBriefs in Mathematics, DOI 10.1007/978-1-4614-5450-2_7,
© Katalin A. Bencsath, Marianna C. Bonanome, Margaret H. Dean, Marcos Zyman 2013

W is finitely generated but not finitely presented. To see why this is the case, recall that a presentation for W is:

$$W = \left\langle a, t;\ \left[a, a^{t^i}\right] = 1\ (i \in \mathbb{Z}) \right\rangle.$$

Suppose that W is finitely presented. Then finitely many of these relators will suffice (Lemma 2.2). Since $W = B \rtimes \langle t \rangle$ then $W/B \cong \langle t \rangle$. Thus, W has an infinite cyclic quotient. Moreover, W does not contain a free subgroup of rank 2 and so, by Theorem 5.1, there exists a finitely generated group A such that $A \leq B$ and either $A^t = B$ or $A^{t^{-1}} = B$. In either case, B would be finitely generated, a contradiction. Hence, W is not finitely presented.

To produce a finitely presented ascending HNN-extension of W, consider the subgroup of W generated by aa^t and t. Put

$$b = b_0 = aa^t = aa_1,\ b_1 = b_0^t = a_1 a_2,\ \ldots,\ b_n = b_{n-1}^t = a_n a_{n+1}, \ldots.$$

It is easy to verify that the subgroup of W generated by

$$\{\ldots, b_{-2}, b_{-1}, b_0, b_1, b_2, \ldots\}$$

is free abelian on these generators, and $gp(aa^t, t) = \langle b \rangle \wr \langle t \rangle$. In particular, $W \cong gp(aa^t, t)$ under the endomorphism taking $a \mapsto b = aa^t$ and $t \mapsto t$. Since $a \notin gp(aa^t, t)$, $gp(aa^t, t) \cong W$ is a proper subgroup of W.

We introduce a stable letter s and form the ascending HNN-extension

$$E = \left\langle W, s;\ a^s = aa^t, t^s = t \right\rangle.$$

Observe that

$$E = gp(a, s, t)$$

and that all the relators for W are so far still present in E. Our aim is to show that in addition to E being finitely generated, it is also finitely presented and metabelian. To achieve this, we think of W in the same way as in Example 3.4.

Let $M = \mathbb{Z}[x, x^{-1}]^+$ be the additive group of the polynomial ring over $\mathbb{Z}$ on x and x^{-1}. M is an additive free abelian group, freely generated by $\{\ldots, x^{-2}, x^{-1}, 1, x, x^2, \ldots\}$. Multiplication by x defines an infinite-order automorphism α of M: $x^i \alpha = x^{i+1}$.

Consider the (external) semi-direct product

$$\widehat{W} = M \rtimes_\varphi \langle \hat{t} \rangle,$$

where $\langle \hat{t} \rangle$ is the infinite cyclic group generated by $\hat{t}$ and $\varphi : \hat{t} \to \alpha$. Observe that $\widehat{W} = gp(1, \hat{t})$ and $\widehat{W} \cong W$ under the isomorphism induced by $1 \mapsto a$ and $\hat{t} \mapsto t$. This isomorphism maps each x^i to a_i. Thus, in particular, $\widehat{W}$ realizes once again the standard wreath product of an infinite cyclic group by another.

Next, consider the ring

$$\mathbb{Z}\left[x, x^{-1}, (1+x)^{-1}\right]$$

obtained by attaching the formal inverse of $(1+x)$ to $\mathbb{Z}[x, x^{-1}]$, and denote the additive group of this ring by $\mathbb{Z}[x, x^{-1}, (1+x)^{-1}]^+$. Let

$$\widehat{E} = \mathbb{Z}[x, x^{-1}, (1+x)^{-1}]^+ \rtimes \langle \hat{t}, \hat{s}; [\hat{s}, \hat{t}] = 1 \rangle,$$

the semi-direct product of $\mathbb{Z}[x, x^{-1}, (1+x)^{-1}]^+$ by the free abelian group $\langle \hat{t}, \hat{s}; [\hat{t}, \hat{s}] = 1 \rangle$. In this semi-direct product, the action of $\hat{t}$ is multiplication by x and the action of $\hat{s}$ is multiplication by $(1+x)$. Therefore,

$$\widehat{E} = gp(1, \hat{s}, \hat{t}).$$

We have constructed a group $\widehat{E}$ such that $E \cong \widehat{E}$ via the mapping $a \mapsto 1, t \mapsto \hat{t}$, and $s \mapsto \hat{s}$. Under this isomorphism,

$$aa^t = a^s \mapsto 1^{\hat{s}} = 1(1+x) = 1+x.$$

The fact that $\widehat{E}$ is a semi-direct product of an abelian group by an abelian group shows that E is metabelian. A presentation for E is

$$E = \left\langle a, t, s; [s,t] = 1, \left[a, a^{t^m}\right] = 1 \ (m \in \mathbb{Z}), a^s = aa^t \right\rangle,$$

which has infinitely many relations. However, the infinitely many commutator relations can be deduced from a single relation as follows:

$$
\begin{aligned}
1 = \left[a, a^t\right] &= \left[a, a^t\right]^s \\
&= \left[a^s, a^{ts}\right] \\
&= \left[aa^t, a^{st}\right] \\
&= \left[aa^t, a^t a^{t^2}\right] \\
&= \left[a, a^{t^2}\right].
\end{aligned}
$$

By iteration, $[a, a^{t^n}] = 1$ for all positive integers n. If $n < 0$, $[a, a^{t^n}]^{t^{-n}} = [a^{t^{-n}}, a]$. Thus, using Tietze transformations (Theorem 2.1), E has the finite presentation:

$$E = \left\langle a, t, s; [s,t] = 1, [a, a^t] = 1, a^s = aa^t \right\rangle,$$

and W has been embedded in a finitely presented metabelian group.

We now prove four lemmas, which will play an important part in the proof of the embedding theorem.

Lemma 7.1. *Every finitely generated metabelian group can be embedded in a finitely generated metabelian group which is a semi-direct product of an abelian group A by a finitely generated abelian group Q. Furthermore, A is a finitely generated $\mathbb{Z}Q$-module.*

Proof. Let G be a finitely generated metabelian group. We can make the short exact sequence

$$1 \longrightarrow G' \longrightarrow G \longrightarrow Q \longrightarrow 1,$$

where $Q = G/G'$, a finitely generated abelian group. By Theorem 4.3, G embeds in $(G')^Q \rtimes Q = G' \wr Q$, the unrestricted wreath product of G' with Q.

Suppose $G = gp(x_1, \ldots, x_k)$, and let $\varphi : G \hookrightarrow (G')^Q \rtimes Q$. Then

$$G\varphi = gp(x_1\varphi, \ldots, x_k\varphi) = gp(u_1 b_1, \ldots, u_k b_k),$$

where $b_i \in (G')^Q$, $u_i \in Q$.

Put $B = gp(b_1, \ldots, b_k, Q) = gp_B(b_1, \ldots, b_k) \rtimes Q$. Let $A = gp_B(b_1, \ldots, b_k)$. Observe that $G\varphi \leq B$, that B is finitely generated (since Q is), and that B is a semi-direct product of the abelian group A by the finitely generated abelian group Q. Since A is finitely generated as a normal subgroup of B, it is a finitely generated $\mathbb{Z}Q$-module, by Lemma 6.3 □

Recall that every submodule of a finitely generated $\mathbb{Z}Q$-module is finitely generated as a module (see Chap. 6).

Lemma 7.2. *Let M be a finitely generated $\mathbb{Z}Q$-module, where Q is a finitely generated abelian group, containing an element $t \in Q$ of infinite order in Q. Then there exists a bimonic polynomial $f = 1 + c_1 t + c_2 t^2 + \cdots + t^n$ $(c_i \in \mathbb{Z})$ such that $mf = 0$ if and only if $m = 0$ $(m \in M)$.*

Proof. Put $M_1 = \{m \in M \mid mf = 0 \text{ for some bimonic polynomial } f\}$. M_1 is in fact a submodule of M and hence it is finitely generated, say by $\{z_1, \ldots, z_k\}$. Let f_i be a bimonic polynomial such that $z_i f_i = 0$ $(i = 1, \ldots, k)$. Notice that if $m \in M_1$, then $mf_1 f_2 \cdots f_k = 0$. Put $f = 1 + f_1 f_2 \cdots f_k$. Of course, we need not be concerned with elements $m \notin M_1$. For $m \in M_1$, $mf = 0$ yields

$$0 = mf = m + 0 = m,$$

as required. □

The ideas contained in Lemma 7.3 come from commutative algebra. We recommend that the reader consult [1].

Lemma 7.3. *Let R be a commutative unital ring. Let S be a submonoid of R; i.e., a multiplicatively closed subset of R containing 1. In addition, assume that $0 \notin S$, and S contains no zero divisors of R. Then R can be expanded to a commutative unitary ring R_S in which the elements of S all have inverses. Furthermore, if M is an R-module and S acts faithfully on M, M can be embedded in an R_S-module.*

Proof. We want to invert all the elements of S. Form a set of formal symbols

$$\frac{R}{S} = \left\{ \frac{r}{s} \,\middle|\, r \in R, s \in S \right\} \cong R \times S \text{ setwise.}$$

Define an equivalence relation $\sim$ on $\dfrac{R}{S}$:

$$\frac{r}{s} \sim \frac{r'}{s'} \text{ if and only if } rs' - sr' = 0.$$

Denote the set of equivalence classes by R_S and the equivalence class of $\frac{r}{s}$ by $\left[\frac{r}{s}\right]$. Since $0 \notin S$ and S has no zero divisors, the mapping $r \mapsto \left[\frac{r}{1}\right]$ $(r \in R)$ is a ring monomorphism.

Define multiplication in R_S by

$$\left[\frac{r_1}{s_1}\right]\left[\frac{r_2}{s_2}\right] = \left[\frac{r_1 r_2}{s_1 s_2}\right]$$

and addition by

$$\left[\frac{r_1}{s_1}\right] + \left[\frac{r_2}{s_2}\right] = \left[\frac{r_1 s_2 + r_2 s_1}{s_1 s_2}\right].$$

If we now identify R with its image in R_S, we find that

$$s\left[\frac{1}{s}\right] = \left[\frac{s}{1}\right]\left[\frac{1}{s}\right] = \left[\frac{s}{s}\right] = \left[\frac{1}{1}\right] = 1.$$

Thus, we denote $\left[\frac{1}{s}\right]$ by s^{-1}. We have now enlarged R to a ring in which all of the elements of S are invertible.

Now let M be an R-module, and S a submonoid of R as in the hypothesis. Assume further that S acts faithfully on M. That is, $ms = 0 \Rightarrow m = 0$ for all $m \in M$ and all $s \in S$. Form the set

$$\frac{M}{S} = \left\{ \frac{m}{s} \,\middle|\, m \in M, s \in S \right\}.$$

Define an equivalence relation $\sim$ on $\dfrac{M}{S}$ as follows:

$$\frac{m}{s} \sim \frac{m'}{s'} \text{ if and only if } ms' - m's = 0.$$

Denote the set of equivalence classes by M_S and the equivalence class of $\dfrac{m}{s}$ by $\left[\dfrac{m}{s}\right]$. M_S becomes an R_S-module by setting

$$\left[\frac{m}{s}\right]\left[\frac{r}{t}\right] = \left[\frac{mr}{st}\right] \quad \left(\left[\frac{m}{s}\right] \in M_S, \left[\frac{r}{t}\right] \in R_S\right).$$

Since S acts faithfully on M, the mapping $m \to \begin{bmatrix} m \\ 1 \end{bmatrix}$ $(m \in M)$ is monic. Thus, M embeds in the R_S-module M_S, as required. $\qquad\qquad\qquad\qquad\qquad$ $\square$

Lemma 7.4. *Let a, b, t, u be elements of a group G and let d be a positive integer. Suppose that*

$$[a^v, b^w] = 1 \text{ whenever } v, w \in \{t^i \mid i = 0, 1, \ldots, d\}$$

and that

$$[t, u] = 1.$$

In addition, suppose that

$$a^u = a^f, b^u = b^f$$

where f is a specific bimonic polynomial in t of degree d in the group ring $\mathbb{Z}G$. Then

$$[a^v, b^w] = 1 \text{ whenever } v, w \in gp(t).$$

Proof. We begin by proving the following, using induction on m, where m is any nonnegative integer:

$$[a^v, b^w] = 1 \text{ whenever } v, w \in \{t^i \mid i = 0, 1, \ldots, d + m\}.$$

For $m = 0$, the statement is true, since that is the original hypothesis. Assume now that the statement is true for all $k < m$. We must show that the statement is therefore true for m.

Suppose that $f = 1 + c_1 t + \cdots + t^d$, and note that $f \cdot t^m$ is a polynomial in t of degree $m + d$. Let $h = (f - t^d)t^m$, a polynomial in t of degree at most $d + m - 1$.

Note also that the hypothesis $a^u = a^f, b^u = b^f$ implies that a commutes with ▰ conjugates of a by certain powers of t, up to and including t^d; likewise b commutes with conjugates of b by these same powers of t.

Using commutator identities (see Sect. 1.1):

$$1 = \left[a, b^{t^m}\right] \qquad \text{(by the induction hypothesis and the fact that } d \geq 1\text{)}$$

$$= \left[a, b^{t^m}\right]^u$$

$$= \left[a^u, b^{ut^m}\right] \qquad \text{(by 1.1 and the fact that } [u, t] = 1\text{)}$$

$$= \left[a^f, b^{f \cdot t^m}\right]$$

$$= \left[a^f, b^{t^{m+d}}\right] \left[a^f, b^h\right]^{b^{t^{m+d}}} \qquad \text{(by 1.3)}$$

$$= \left[a^f, b^{t^{m+d}} \right] \qquad \text{(by the induction hypothesis)}$$

$$= \left[a \cdot a^{c_1 t + \cdots + t^d}, b^{t^{m+d}} \right]$$

$$= \left[a^{c_1 t + \cdots + t^d}, b^{t^{m+d}} \right]^a \left[a, b^{t^{m+d}} \right] \qquad \text{(by 1.2)}$$

$$= \left[a^{c_1 + \cdots + t^{d-1}}, b^{t^{m+d-1}} \right]^{ta} \left[a, b^{t^{m+d}} \right] \qquad \text{(by 1.1)}$$

$$= \left[a, b^{t^{m+d}} \right] \qquad \text{(by the induction hypothesis).}$$

Thus, $[a, b^{t^{m+d}}] = 1$, and similar reasoning shows that $[a^{t^{m+d}}, b] = 1$, for $m = 0, 1, \ldots$.

Now consider $[a^{t^n}, b^{t^p}]$ $(n, p \in \mathbb{Z}^+)$. Without loss of generality, assume $n < p$. Then

$$\left[a^{t^n}, b^{t^p} \right] = \left[a, b^{t^{p-n}} \right]^{t^n} = 1.$$

Computation with non-positive powers of t follows easily in the same way. □

Example 7.1 provides enough motivation for the more general theorem, indeed the high point of this chapter.

Theorem 7.1. *If G is a finitely generated metabelian group, then G can be embedded in a finitely presented metabelian group.*

Proof. Let G be a finitely generated metabelian group. By Lemma 7.1, there is an abelian group A and a finitely generated abelian group Q such that G embeds in $B = A \rtimes Q$.

Since A is finitely generated as a $\mathbb{Z}Q$-module, let $\{b_1, \ldots, b_r\}$ be a set of module generators for A. Since $\mathbb{Z}Q$ is a Noetherian ring, every submodule of the finitely generated $\mathbb{Z}Q$-module A is itself finitely generated. In particular, if T is the torsion subgroup of A, then T is a finitely generated $\mathbb{Z}Q$-submodule of A; generated, say, by $\{a_1, \ldots, a_m\}$, where a_i is of order $\beta_i > 0$ for $i = 1, \ldots, m$. Every torsion element of A is a product of conjugates of a_i's (not necessarily distinct) by elements of Q; hence, its order is a consequence of the relations $a_i^{\beta_i} = 1$.

If Q is finite, B is finitely presented and we are done because B is metabelian. Assume Q has at least one element of infinite order. Let

$$\{t_1, \ldots, t_k, t_{k+1}, \ldots, t_l\}$$

be a basis for the finitely generated abelian group Q; where t_i is of infinite order for $i = 1, \ldots, k$; and t_i is of order $e_i > 0$ for $i = k+1, \ldots, l$.

By Lemma 7.2, there are positive integers d_i and corresponding bimonic polynomials

$$f_i = 1 + c_{i1}t_i + c_{i2}t_i^2 \cdots + t_i^{d_i} \in \mathbb{Z}Q \ (i = 1, \ldots, k)$$

such that

$$a^{f_i} = 1 \text{ if and only if } a = 1.$$

Thus, $F = \{f_1, \ldots, f_k\}$ acts faithfully on A.

We now concoct a group E_F, dependent on both the group B and the set F. E_F will turn out to be an *HNN*-extension with base group B. For this reason, several of the generating elements of E_F have the same name as the generating elements of B, and many of the defining relations are dictated by the defining relations of B.

Let the generating set for E_F be

$$\{b_1, \ldots, b_r, t_1, \ldots, t_l, u_1, \ldots, u_k\}.$$

The defining relations for E_F fall into eight types:

1. $[b_i, b_j] = 1 \quad (i, j = 1, \ldots, r) \quad$ (A is abelian)
2. $[t_i, t_j] = 1 \quad (i, j = 1, \ldots, l) \quad$ (Q is abelian)
3. $t_i^{e_i} = 1 \quad (i = k+1, \ldots, l) \quad$ (the torsion elements of Q)
4. $[b_i^v, b_j^w] = 1 \quad (i, j = 1, \ldots r)$, where
$$v, w \in \{t_1^{\alpha_1} \cdots t_l^{\alpha_l} \mid 0 \le \alpha_j \le d_j \quad (j = 1, \ldots, k) \text{ and }$$
$$0 \le \alpha_j \le e_j \quad (j = k+1, \ldots, l)\} \quad \text{(A is abelian)}$$
5. $b_i^{u_j} = b_i^{f_j} \quad (i = 1, \ldots, r; \ j = 1, \ldots, k) \quad$ (the action of the u_j's on the b_i's)
6. $[u_i, u_j] = 1 \quad (i, j = 1, \ldots, k) \quad$ (the u_i's commute)
7. $t_j^{u_i} = t_j \quad (i = 1, \ldots, k; \ j = 1, \ldots, l) \quad$ (the u_i's act trivially on the t_j's)
8. $a_i^{\beta_i} = 1 \quad (i = 1, \ldots, m) \quad$ (the torsion in A)

Note that E_F is finitely presented. Our goal is to show that E_F contains a copy of B and that E_F is metabelian.

Now in B, an infinite number of relations of the form

$$[b_i^v, b_j^w] = 1 \quad \left(v, w \in \{t_1^{\alpha_1} \cdots t_l^{\alpha_l} \mid \alpha_j \in \mathbb{Z} \ (j = 1, \ldots, k)\right.$$
$$\left. \text{and } 0 \le \alpha_j \le e_j \ (j = k+1, \ldots, l)\}\right)$$

are required in order to encode that A is abelian. However, in E_F, by Lemma 7.4, each of the relations given in 7 is a consequence of the finitely many relations from (4) above. Hence, we see that E_F is actually an (external) iterated ascending *HNN*-extension of B with k commuting stable letters added (see Example 3.10):

$$E_F = \left\langle B, u_1, \ldots, u_k; \ b_i^{u_j} = b_i^{f_j} \ (i = 1, \ldots, r; \ j = 1, \ldots, k), \right.$$
$$t_i^{u_j} = t_i \ (i = 1, \ldots, l; \ j = 1, \ldots, k),$$
$$\left. [u_i, u_j] = 1 \ (i, j = 1, \ldots, k) \right\rangle.$$

Hence, B embeds in E_F (see Sect. 3.3.2).

To see that E_F is metabelian, consider the set of bimonic polynomials F given in 7, and expand the group ring $\mathbb{Z}Q$ to a ring $\mathbb{Z}Q_F$ where every element of F is a unit.

Then consider the $\mathbb{Z}Q_F$-module A_F as in Lemma 7.3. A_F is in fact finitely generated as a $\mathbb{Z}Q_F$-module by the set $\{b_1,\ldots,b_r\}$.

Let $\widehat{U}$ be a free abelian group with basis

$$\{\hat{u}_1,\ldots,\hat{u}_k\},$$

and consider the direct product:

$$\widehat{P} = Q \times \widehat{U}.$$

Let

$$\widehat{E}_F = A_F \rtimes \widehat{P},$$

where Q acts on the b_i's by conjugation, and $\widehat{U}$ acts on A_F as follows:

$$b_i^{\hat{u}_j} = b_i^{f_j} \ (i = 1,\ldots,r, \ j = 1,\ldots,k).$$

We have constructed a group $\widehat{E}_F$ such that $E_F \cong \widehat{E}_F$ via the mapping $b_i \mapsto b_i$ $(i = 1, \ldots,r)$, $t_i \mapsto t_i$ $(i = 1,\ldots,l)$, $u_i \mapsto \hat{u}_i$ $(i = 1,\ldots,k)$.

The fact that $\widehat{E}_F$ is a semi-direct product of an abelian group by an abelian group shows that E_F is metabelian. $\square$

Appendix A
Sketch of Proof of Lemma 1.1

It is helpful to compare this matrix group to the two generator 3-solvable group originally constructed by Hall in [8]. This construction was also reviewed recently by Baumslag et. al. in [6].

Recall that $G = gp(A, B)$, where

$$A = \begin{pmatrix} 1 & 0 & 0 \\ 0 & x & 0 \\ 0 & 0 & 1 \end{pmatrix} \quad \text{and} \quad B = \begin{pmatrix} 1 & 0 & 0 \\ 1 & 1 & 0 \\ 0 & 1 & 1 \end{pmatrix},$$

with entries from $\mathbb{Q}(x)$, the quotient field of the ring of polynomials $\mathbb{Q}[x]$ over the rational numbers. Proposition 1.1 asserts that the center of G is a free abelian group of countably infinite rank.

It is a good idea to actually work through the details in the steps below outlining the proof.

Proof (Sketch).

I. Begin by showing that $G/\zeta(G) \cong \mathbb{Z} \wr \mathbb{Z}$. To that end:

(a) Show that the elements of G are matrices all of whose entries come from $\mathbb{Z}[x, x^{-1}]$.

(b) Put $H = gp_G(B)$.
   Show that a typical element $h \in H$ has the form:

$$h = \begin{pmatrix} 1 & 0 & 0 \\ f(x) & 1 & 0 \\ * & f(x^{-1}) & 1 \end{pmatrix}$$

where $f(x)$ can be any element in $\mathbb{Z}[x, x^{-1}]$ and $* \in \mathbb{Z}[x, x^{-1}]$ (The reader may wish to try to find restrictions, if any, on $*$.).

K.A. Bencsáth et al., *Lectures on Finitely Generated Solvable Groups*,
SpringerBriefs in Mathematics, DOI 10.1007/978-1-4614-5450-2,
© Katalin A. Bencsath, Marianna C. Bonanome, Margaret H. Dean, Marcos Zyman 2013

(c) Observe that if $h \in H$, then

$$A^{-1}hA = h^A = \begin{pmatrix} 1 & 0 & 0 \\ xf(x) & 1 & 0 \\ ** & x^{-1}f(x^{-1}) & 1 \end{pmatrix} \tag{A.1}$$

with $** \in \mathbb{Z}[x, x^{-1}]$. In particular, this indicates that $A \notin \zeta(G)$.
   It follows that $G = H \rtimes gp(A)$:

   (i) $G = gp(A)H$ (given)
   (ii) $gp(A) \cap H = I$

(d) Now, it is easy to see that $\zeta(G) \leq H$. Thus,

$$G/\zeta(G) = H/\zeta(G) \rtimes gp(A),$$

with $gp(A)$ an infinite cyclic group. Note that $G/\zeta(G) = gp(\bar{A}, \bar{B})$, where
$\bar{A} = A\zeta(G), \bar{B} = B\zeta(G)$ in $G/\zeta(G)$.

(e) Modulo the center, elements h_1 and h_2 are equivalent if and only if they
differ at most in the lower left corner entry only. Note that $\bar{h}_i\bar{h}_j$ is additive in
the off diagonal, which suggests that we can put the equivalence classes of
$H/\zeta(G)$ in a one-to-one correspondence with $\{f(x)|f(x) \in \mathbb{Z}[x, x^{-1}]\}$. This
immediately implies that $H/\zeta(G) \cong \mathbb{Z}[x, x^{-1}]$ viewed as an additive group.
Thus, we have

$$G/\zeta(G) \cong \mathbb{Z}[x, x^{-1}] \rtimes gp(A) \cong \mathbb{Z} \wr \mathbb{Z} = \langle a, b; [b, b^{a^i}] = 1 \ (i \in \mathbb{Z}) \rangle$$

(see Example 3.5).

II. Show that $\zeta(G)$ is free abelian of countably infinite rank. To that end:

(a) Let ϕ be the canonical homomorphism of G onto its quotient group $G/\zeta(G)$
and η the isomorphism

$$\eta : G/\zeta(G) \to \mathbb{Z} \wr \mathbb{Z} = \langle a, b; [b, b^{a^i}] = 1 \ (i \in \mathbb{Z}) \rangle$$

described by

$$\eta : \begin{cases} \bar{B} & \mapsto b \\ \bar{A} & \mapsto a \end{cases}.$$

Then the composition $\phi\eta = \theta$ gives us an epimorphism

$$\theta : G \to \langle a, b; [b, b^{a^i}] = 1 \ (i \in \mathbb{Z}) \rangle$$

with $ker(\theta) = \zeta(G)$.

(b) But $ker(\theta) = gp_G([B,B^{A^i}])$. So it follows that $\zeta(G) = H'$ since

$$H' = gp_H(\{[B,B^{A^i}]\}\ (i \in \mathbb{Z})) = gp_G(\{[B,B^{A^i}]\}\ (i \in \mathbb{Z})).$$

Further tedious calculations show that

$$H' = \zeta(G) = \left\{ \begin{pmatrix} 1 & 0 & 0 \\ 0 & 1 & 0 \\ k & 0 & 1 \end{pmatrix} : \exists f,g \in \mathbb{Z}[x,x^{-1}],\ k = f(x)g(x^{-1}) - f(x^{-1})g(x) \right\}.$$

$\square$

Appendix B
Theorem 2.1 Details

Proof. Recall that if $G = \langle X; R \rangle$, then there exists a map θ from X to G whose extension to $F = F(X)$, the free group on X, gives a homomorphism with kernel $gp_F(R)$. **T2** and **T2′** follow from observing that if $S \subseteq gp_F(R)$, then

$$gp_F(R) = gp_F(R \cup S).$$

To prove **T1**, we argue as follows. Let Y be disjoint from X and let

$$\{w_y(\tilde{x}) : y \in Y\}$$

be a set of X-words indexed by the elements of Y (where for each y, $\tilde{x}$ denotes the list of elements of X that appear in w_y.). Consider the presentation map $\theta : X \to G$ and define

$$\theta^+ : X \cup Y \to G$$

in the following way: $\theta^+ = \theta$ on X, and $y\theta^+ = w_y(\tilde{x}\theta)$ for each $y \in Y$. Thus, for each $y \in Y$, $y\theta^+$ is the element of G obtained by replacing each occurrence of x in w_y by $x\theta$.

Let F^+ be the free group on $X \cup Y$. It is clear that

$$\theta^+ : X \cup Y \to G$$

extends to a surjective homomorphism from F^+ to G and its kernel is

$$gp_{F^+}\left(R \cup \{y^{-1}w_y(\tilde{x}) : y \in Y\}\right).$$

Next, let

$$\tilde{Y} = \{y^{-1}w_y(\tilde{x}) : y \in Y\}$$

and notice that F^+ is free on $X \cup \tilde{Y}$. Since θ^+ sends x to $x\theta$ and each $y^{-1}w_y(\tilde{x})$ to 1, this can be regarded as the definition of θ^+. So the claim is verified. The proof of **T1′** is entirely analogous. $\qquad\square$

K.A. Bencsáth et al., *Lectures on Finitely Generated Solvable Groups*,
SpringerBriefs in Mathematics, DOI 10.1007/978-1-4614-5450-2,
© Katalin A. Bencsath, Marianna C. Bonanome, Margaret H. Dean, Marcos Zyman 2013

Appendix C
Presenting an (Internal) HNN-Extension

Lemma C.1. *Suppose E is an HNN-extension of B, with stable letter t, associating isomorphism φ and associated subgroups H and K. If $B = \langle X; R \rangle$, then $E = \langle X, t; R \cup \{ t^{-1} u_i t = v_i \mid i \in I \} \rangle$, where $H = gp(u_i \mid i \in I)$, $K = gp(v_i \mid i \in I)$, and $v_i = u_i \varphi$ $(i \in I)$.*

(This assumes that given an isomorphism φ between H, K, we can find generating sets for H, K, precisely corresponding to each other under this isomorphism φ.)

Proof. Recall that presentations come with hidden maps floating around (see Chap. 2). Let F be free on X and α the presentation map $\alpha : X \to B$ such that α induces a homomorphism α_* from F onto B such that $ker \alpha_* = gp_F(R)$.

We claim that if F^+ is free on $X \cup \{t\}$ and α^+ is the map from $X \cup \{t\}$ to E which is α on X and sends t to t (t is schizophrenic), then the extension α_*^+ from F^+ to E has kernel $gp_{F^+}\left(R \cup \{ t^{-1} u_i t v_i^{-1} \mid i \in I \} \right)$.

Denote $gp_{F^+}\left(R \cup \{ t^{-1} u_i t v_i^{-1} \mid i \in I \} \right)$ by L. If $K_* = ker \alpha_*^+$, then clearly $K_* \supseteq L$. In fact, we claim that $K_* = L$.

To see this, we will work in F^+/L. Let us look at the elements of F^+/L.

Put $gp(X)L/L = B_*$. Then we have the short exact sequence

$$1 \longrightarrow K_* \longrightarrow F^+ \longrightarrow E \longrightarrow 1.$$

There exists an obvious pre-image of B (B is, of course, generated by the image of X under α).

Put

$$x_* = xL,$$

$$B_* = gp(x_* \mid x \in X),$$

$$t_* = tL.$$

K.A. Bencsáth et al., *Lectures on Finitely Generated Solvable Groups*,
SpringerBriefs in Mathematics, DOI 10.1007/978-1-4614-5450-2,
© Katalin A. Bencsath, Marianna C. Bonanome, Margaret H. Dean, Marcos Zyman 2013

In $F^+/L \cong E$, we have produced a set of generators; $F^+/L = gp(B_*, t_*)$. So, every element can be written as a B_*-word with t_*-powers interspersed: $t_*^{\varepsilon_1} b_{*1} t_*^{\varepsilon_2} \cdots t_*^{\varepsilon_n} b_{*n}$. How can an element of F^+/L get mapped to 1 when F^+ is getting mapped to E? t_*^m maps to $t^m \neq 1$. b_* maps to 1 only if it is in $gp_{F^+}(X)$; i.e., in L already. α_*^+ sends the alternating product to $t^{\varepsilon_1} b_1 \cdots t^{\varepsilon_n} b_n$, that can go to 1 only if there is a pinch. So only those elements of L can go to 1. Hence, $K = L$. $\square$

References

1. Atiyah, M.F., Macdonald, I.G.: Introduction to Commutative Algebra. Addison-Wesley, Reading (1969)
2. Baumslag, G.: On finitely presented metabelian groups. Bull. Amer. Math. Soc. **78**(2), 279 (1972)
3. Baumslag, G.: Subgroups of finitely presented metabelian groups. J. Austral. Math. Soc. **16**, 98–110 (1973)
4. Baumslag, G., Cannonito, F.B., Miller, C.F., III: Infinitely generated subgroups of finitely presented groups. II. Math. Z. **172**(2), 97–105 (1980)
5. Baumslag, G., Cannonito, F.B., Robinson, D.J.S.: The algorithmic theory of finitely generated metabelian groups. Tran. Amer. Math. Soc. **344**(2), 629–648 (1994)
6. Baumslag, G., Mikhailov, R., Orr, K.E.: A new look at finitely generated metabelian groups, arXiv:1203.5431v1 [math.GR] 24 Mar 2012
7. Bieri, R., Strebel, R.: Almost finitely presented soluble groups. Comment. Math. Helv. **53**, 258–278 (1978)
8. Hall, P.: Finiteness conditions for soluble groups. Proc. London Math. Soc. **4**, 419–436 (1954)
9. Hall, P.: On the finiteness of certain soluble groups. Proc. London Math. Soc. **3**, 595–622 (1959)
10. Hall, P.: The Frattini subgroups of finitely generated groups. Proc. London Math. Soc. **3**, 327–352 (1961)
11. Higman, G.: Subgroups of finitely presented groups. Proc. Roy. Soc. Ser. A **262**, 455–475 (1961)
12. Krasner, M., Kaloujnine, L.: Produit complet des groupes de permutations et probleme dextension des groupes. Acta Sci. Mz. Szeged **13**, 208–230 (1950); **14**, 39–66 and 69–82 (1951)